# *First Steps in LaTeX*

George Grätzer

# *First Steps in LaTeX*

BIRKHÄUSER • SPRINGER
BOSTON NEW YORK

George Grätzer
Department of Mathematics
University of Manitoba
Winnipeg, MB R3T 2N2
Canada

**Library of Congress Cataloging-In-Publication Data**

Gratzer, George A.
First steps in LaTeX / George Gratzer.
p. cm.
Includes bibliographical references (p. ) and index.
ISBN 0-8176-4132-7 (alk. paper)
1. LaTeX (computer file) 2. Computerized typesetting. I. Title.
Z253.4.L38 G74 1999 99-32249
686.2'25445369–dc21 CIP

Printed on acid-free paper


ISBN 0-8176-4132-7 SPIN #19901659
ISBN 3-7643-4132-7

Typeset by the author in LaTeX.
Printed and bound by Hamilton Printing, Rensselaer, NY.
Printed in the United States of America.

9 8 7 6 5 4 3 2 1

# *Contents*

# Quick Finder

# *Introduction*

## *Are you in a hurry?*

A friend received a letter from the American Mathematical Society (AMS) informing him that his paper had been accepted for publication in the *Proceedings of the AMS*. If he submitted it as a LaTeX document, it would be published in 20 weeks—any other format would take almost a year before the appearance in print of the article.

The friend had LaTeX installed on his computer on Friday, borrowed the manuscript of this book, and mailed a LaTeX version of his article to the AMS on Monday.

*First Steps in LaTeX* is for the mathematician, physicist, engineer, scientist, or technical typist who needs to quickly learn how to write and typeset articles containing mathematical formulas.

A quick introduction to LaTeX and the AMS enhancements is provided so that you will be ready to prepare your first article (such as the sample articles on pages 53–54 and 67–69) in only a few hours.

Specific topics can be found in the table of contents, the Quick Finder, or the index. While the index is LaTeX-oriented, the Quick Finder lists the main topics using terminology common to wordprocessing applications. For example, to find out how to italicize text, look under *italics* in the Quick Finder.

## *Setting the stage*

Watch someone type a mathematical article in LaTeX. You will see how to

- *Type the document* using a *text editor* to create a LaTeX *source file.*
  A *source file* might look like this (we will refer to this as `first.tex`):

```
\documentclass{article}
\begin{document}
The hypotenuse: $\sqrt{a^{2} + b^{2}}$.  I can type math!
\end{document}
```

`first.tex` is different from a wordprocessing file: All letters are displayed at the same size and in the same font.

- *Typeset the file and view it on the monitor.* Typesetting gives the following (the two corners are hints that the material was typeset by LaTeX):

The hypotenuse: $\sqrt{a^2 + b^2}$. I can type math!

- *Continue the editing cycle.* The typist may go back and forth between the source file and the typeset version, noting that as changes are made, they are quickly reflected in the typeset version.
- *Print the file.* Once the typeset document is satisfactory, the article may be printed to provide a paper version of the typeset article.

Unfortunately, one cannot tell exactly how any particular text editor works, or how the typesetting and printing is done on your system. Just as there are many text editors (ranging from the ancient `vi` to modern editors with graphical user interfaces), there are many LaTeX setups, each with its own unique installation and a different way of typesetting and printing. However, the following two examples should give you some idea of the process.

***Example 1:* UNIX**

UNIX commands are typed at a *shell prompt* (`unix$`). Type the following command to start a text editor:

```
unix$ vi first.tex
```

Once the editor is started, the text of the article may be typed. When the file is complete, save the file and exit the editor. The article is now ready to be typeset. Back at the shell prompt, type:

```
unix$ latex first
```

which results in a series of messages scrolling up the monitor as the file is typeset.

When this process is completed, one has a DVI file, `first.dvi`, that can be viewed (in an X Windows environment) by typing

```
unix$ xdvi first
```

If more changes are required, the text editor can be restarted—editing, saving, quitting, and typesetting the article again.

To print the DVI file, type the following command at a shell prompt:

```
unix$ dvips first | lpr
```

### *Example 2:* TEXTURES *on a Macintosh*

When TEXTURES (on a Macintosh computer) starts up, a blank text-editing window appears. The text of the document is typed in the window and is saved as `first.tex`—the window is now named `first.tex`. When the document is ready to be typeset, select the `LaTeX` format in the `Typeset` menu; then choose `Typeset` from the same menu.

A second window (called `first.tex typeset`) appears, displaying the typeset version of the document (see Figure 1).

If all is satisfactory with the typeset version, choose `Print...` from the `File` menu to print the document. If more editing is required, then simply click the mouse on the `first.tex` window.

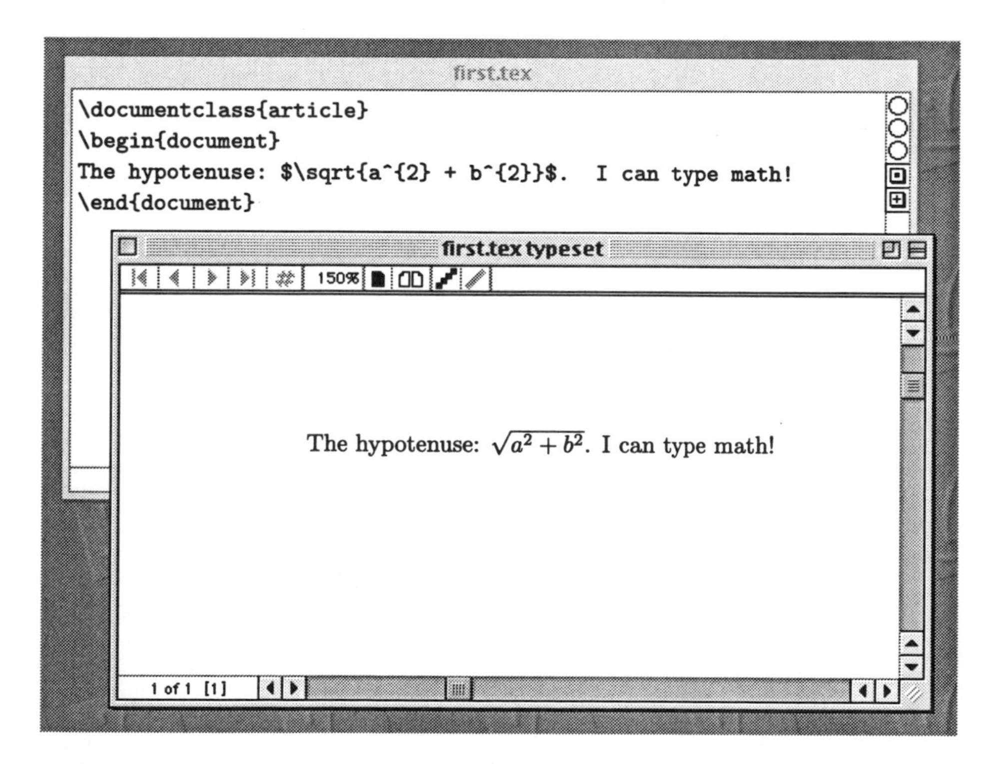


Figure 1: TEXTURES typesetting `first.tex`.

With TEXTURES, changes in the `source file` may be automatically reflected in the typeset version. Go to `Options` in the `Typeset` menu and choose `Flash mode`. Now typesetting takes place as the source file is being typed or edited.

### *Choosing a LaTeX setup*

In Sections C.2 and C.3, we briefly review a few of the most popular LaTeX setups. My best advice to you is to get the same LaTeX setup as that of a friend or colleague so they can help you get started. Many LaTeX implementations come on a CD-ROM or are downloadable from the Internet, and installation is often no more complicated than double-clicking on an icon. Learning the fundamentals of using a text editor (typing text and simple editing) is easy—if you know how to use a word processor, you already know how to use a text editor. And it is equally easy to become familiar with a few basic commands (for tasks such as typesetting and printing).

## *What is document markup?*

Working with a word processor, the document is visible on the monitor, more or less as it will look in print, with different fonts, font sizes, font shapes (e.g., roman, italic) and weights (e.g., normal, boldface), interline spacing (leading), indentation, and so on.

Working with a *markup language* is different: Text is entered into a *source file* that is visible on the monitor with all letters having the same font, the same size and shape. To indicate changes in the appearance of the typeset text, one must "mark up" the source file, that is, add commands to format the text. For instance, to emphasize the phrase "`detailed description`" in a LaTeX source file, type

```
\emph{detailed description}
```

The `\emph` command is a markup command, which will result in the typeset output:

*detailed description*

In this small book, you will be introduced to LaTeX, a markup language designed and implemented by Leslie Lamport,[1] based on Donald E. Knuth's type-

[1] Since Lamport decided not to develop LaTeX any further himself, a talented group of mathematicians and programmers—Frank Mittelbach, Chris Rowley, and Rainer Schöpf—formed the "LaTeX3 team" with the aim of updating, actively supporting, and maintaining LaTeX. The group's membership has since expanded with the addition of Johannes Braams, David Carlisle, Michael Downes, Denys Duchier, and Alan Jeffrey. *This book only discusses the current version of LaTeX*, often called LaTeX $2_\varepsilon$.

setting language TEX (see references [4] and [5] in Section 6.5). You will find that it is quite easy to learn how to mark up text.

On pages 70–77, the source file for `sampart.tex`, the AMS sample article, is juxtaposed with the typeset version. The markup in the source file may appear somewhat challenging at first, but the typeset article is certainly a pleasing rendering of the marked-up material.

# *TEX*

The markup language *TEX* was designed for typesetting mathematical and scientific articles and books, and can handle complex mathematical formulas as well as text: To get the formula $\int_{0}^{\pi} \sqrt{\alpha^{2} + x^{2}}\,dx$, type

```
$\int_{0}^{\pi} \sqrt{\alpha^{2} + x^{2}}\,dx$
```

You do not have to worry about determining the size of the integral symbol or constructing the square root symbol that covers $\alpha^2 + x^2$, because TEX does this for you!

A tremendous part of the appeal of the TEX language is that source files are ***plain text*** (ASCII text), which is easy to transmit ***electronically*** to colleagues, co-authors, journals, editors, and publishers.

TEX is also ***platform independent***. You may type the first version of a source file on a Macintosh computer; your co-author may make improvements to the same file on a PC (a computer running Microsoft Windows); and the journal publishing the article may use a UNIX machine (a computer running a UNIX variant such as Solaris or Linux) to prepare the manuscript for printing.[2]

# *LATEX*

*LATEX* was built on TEX's foundation, and has commands that are easier to use, a set of structural elements, and a larger set of diagnostic messages.

LATEX provides the following additional features:

- An article is divided into ***logical units***, including an abstract, various sections and subsections, theorems, and a bibliography. The logical units are typed independently of one another. Once all the units have been typed, LATEX controls the ***placement*** and ***formatting*** of these elements.

  Line 4 of the sample article `intrart.tex` (on page 48) reads
  `\documentclass{article}`

[2]You may take this sentence as a definition of the three major computing platforms. ***We will not discuss any tools that are not available in some form on all three platforms.***

This line tells LaTeX to utilize the *document class* `article`, which formats the document as a generic article. When submitting your article to a journal that handles LaTeX manuscripts, the editor can simply substitute the name of the journal's document class to make the body of your article conform to the journal's design (depending on the journal, you may need to use the AMS article document class, `amsart`, in order to ensure that the front matter also adapts). Many journals make their document classes available to prospective authors.

- LaTeX relieves you of tedious bookkeeping chores. Imagine that you have finished writing an article, with all of your theorems and equations numbered and properly cross-referenced. After a final reading, some changes must be made: Section 4 needs to be moved after Section 7, and some new theorems have to be inserted somewhere in the middle. Such a minor change used to be a major headache! But with LaTeX, it almost becomes a pleasure to make such changes. LaTeX automatically renumbers the sections, theorems, and equations in your article, and rebuilds the cross-references.

## *The AMS packages*

The AMS packages are built on top of LaTeX, and enhance LaTeX's capabilities in three different areas:

**1. Math.** The AMS packages provide:

- Powerful tools to deal with multiline math formulas. For instance, in the following formula, the equals signs (=) and the explanatory comments are vertically aligned:

$$\begin{aligned} x &= (x+y)(x+z) && \text{(by distributivity)}\\ &= x+yz && \text{(by Condition (M))}\\ &= yz. \end{aligned}$$

- Numerous constructs for typesetting mathematical formulas, exemplified by the following:

$$f(x) = \begin{cases} -x^2, & \text{if } x < 0;\\ \alpha + x, & \text{if } 0 \leq x \leq 1;\\ x^2, & \text{otherwise.} \end{cases}$$

- The `proof` environment and three theorem styles: plain, definition, and remark. (See the `sampart.tex` sample article on pages 67–69: Theorem 1 uses the plain style, Definition 1 uses the definition style, and the Notation uses the remark style.)

2. **Document classes.** The AMS provides a number of document classes; the most important is the AMS article document class, `amsart`, which allows the input of title page information (e.g., author, address, e-mail address) as separate entities. As a result, a journal can typeset even the title page of an article according to its own specifications without having to retype any information.

3. **Fonts.** Hundreds of mathematical symbols are provided by the AMS packages (see Appendix A). Here are just a few:

$$\leftleftarrows, \quad \blacktriangle, \quad \nexists, \quad \supsetneqq, \quad \mathbb{A}, \quad \mathfrak{p}, \quad \mathcal{E}$$

The AMS calls these enhancements $\mathcal{AMS}$-LaTeX (the math packages and the document classes) and AMSFonts (the font-related packages and the fonts themselves). In this book, to simplify the terminology, we refer to all these enhancements collectively as AMS *packages;* AMS *distribution* and AMS *enhancements* are used as synonyms.

We will point out in the text which commands are LaTeX commands and which are defined by AMS packages. References to AMS commands will also be indicated

by the use of a symbol in the margin (like the one here). A slightly smaller version of this symbol, Ⓐ, is used in the index.

# *What is in the book?*

Just before this introduction is the "Quick Finder," a brief index using mainly non-LaTeX terms.

Chapter 1 shows you how to mark up text (which is quite easy) and Chapter 2 discusses how to mark up math (which is not quite as straightforward). Five sections will ease you into mathematical typesetting, including a discussion of the basic building blocks of formulas.

In Chapter 3, you will practice typing formulas: Section 3.1 is a formula gallery with 20 formulas demonstrating many new concepts. After working through the examples in this section, you will probably notice that these formulas could be typed more efficiently if we had a shorthand for repeated constructs. LaTeX provides such a shorthand, called a *user-defined command,* and this facility is introduced in Section 3.2. Finally, Section 3.3 demonstrates how you can build a complicated formula from a series of simpler formulas.

Chapter 4 describes the anatomy of an article and how to set up an article template. We then walk you through the first sample article, which uses LaTeX's `article` document class.

In Chapter 5, the second sample article, `sampart.tex` is introduced, using `amsart`, the AMS article document class. `sampart.tex` is first shown in typeset form (pages 67–69), then in "mixed" form, showing the source file and the typeset article juxtaposed (pages 70–77). You can learn a great deal about LaTeX and the

AMS packages just by reading the source file one paragraph at a time, and then viewing how that paragraph is typeset.

This introductory book concludes with Chapter 6, which discusses working with LaTeX, including how to interpret LaTeX error messages, the distinction between logical and visual design, spell checkers and text editors, using LaTeX with non-English languages, and further reading about TeX and LaTeX.

You will probably find yourself referring to Appendices A and B quite often: They contain the math and text symbol tables.

Finally, Appendix C briefly discusses TeX, LaTeX, and the Internet. The main topics are:

- Obtaining files from the Internet
- CTAN, the Comprehensive TeX Archive Network
- Obtaining the LaTeX distribution and the AMS packages
- Getting the sample files for this book
- Some commercial TeX implementations
- Freeware and shareware TeX implementations
- TeX user groups and the AMS
- Useful LaTeX information on the Internet
- Sharing your work using the Internet

# *A recommendation*

Use the `amsart` document class for all your articles. Begin each article with the lines: Ⓐ

```
\documentclass{amsart}
\usepackage{amssymb,latexsym}
\begin{document}
```

and ignore all of the discussions in this book about LaTeX commands versus AMS commands, and LaTeX fonts and the latexsym package versus AMS fonts and the amssymb package.

Some of you may not be able to follow this recommendation, including those who work with older installations whose system managers may not install a newer version of LaTeX or the AMS packages, and those who are forced to use a publisher's document class file that is not compatible with the AMS packages. But most users of LaTeX who typeset documents with significant amounts of math will find that using the `amsart` document class and loading amssymb and latexsym make their work easier.

# Keeping up-to-date

Like most computer-related subjects, the material in this book is subject to change over time. While LaTeX itself may not change much until the advent of LaTeX3, there is a new version of the amsmath package on the horizon, introducing a variant of the `equation` environment that will automatically break long formulas into shorter lines. Appendix C deals with the Internet; this information may become obsolete very fast. To keep you up-to-date, I maintain a Web page that will track these changes for you. To find this page, go to my home page:
`http://www.maths.umanitoba.ca/homepages/gratzer/`
and follow the links: `LaTeX books` and `Update`. Or go directly to
`http://www.maths.umanitoba.ca/homepages/gratzer/LaTeXBooks/update.html`

# Conventions

To make this book easy to read, the following simple conventions have been used:

- Explanatory text is set in this typeface: Galliard.
- `The Computer Modern typewriter typeface is used to show what you should type (and also the LaTeX messages). All the characters in this typeface have the same width, making it easy to recognize.`
- The same typeface is also used to indicate
  - Commands (`\emph`)
  - Environments (`align`)
  - Documents (`intrart.tex`)
  - Document classes (`article`)
  - Directories or folders (`work`)
- The names of *packages,* which are extensions of LaTeX, are set in a sans serif typeface (amsmath).
- Computer Modern Roman, which is TeX's standard typeface, is used to display how something looks when typeset:

This typeface—hopefully— looks sufficiently different from the other typefaces used (the strokes are much lighter) that you should not have much difficulty recognizing typeset LaTeX material. When the typeset material is a separate paragraph (or paragraphs), corner brackets in the margin set it off.

- For explanations in the text such as "Compare 'iff' with 'iff', typed as `iff` and `if{f}`, respectively," the same typefaces are used. Because they are not set off visually, it may be a little more difficult to see that 'iff' is set in Computer Modern roman, whereas '`iff`' is set in the Computer Modern typewriter typeface.
- The term *directory* means both directory and folder.

# *Acknowledgments*

This book is based on previous editions of my book, *Math into LATEX* (see Section 6.5). I would like to thank the many people, too numerous to list again here, who read and reread those earlier manuscripts.

My deepest appreciation to all who sent reports on the manuscript of this book.

I received professional reports from Barbara Beeton, Edwin Beschler, David Carlisle, Nandor Sieben, Christina Thiele, and Ferenc Wettl.

A number of volunteers answered my request for help and sent excellent reports with many helpful suggestions:

> Jeffrey Adler, Murray Bell, Charles Blair, Thierry Bouche, Harish Cherukuri, Patrick Cousot, Jean Dezert, David L. Elliott, Luiz Henrique de Figueiredo, Joseph Aloysius Gilvary, Berthold Horn, Bernard Knaepen, Richard Lord, Al Moore, Rafael R. Pappalardo, Dave Pawson, Krzysztof Pszczola, Christopher C. Taylor, Andreas Scherer, Hilmar Schlegel, Stefan Schmiedl, Joseph C. Slater, Boris Veytsman, János Virágh, David Wilson, and Rick Zaccone.

Claire M. Connelly did an outstanding job editing the manuscript, far and beyond the call of duty. Ann Kostant demonstrated that publishers care.

George Grätzer
E-mail: gratzer@cc.umanitoba.ca
Home page: http://www.maths.umanitoba.ca/homepages/gratzer/

CHAPTER

# 1

# *Typing text*

In this chapter, you will learn how easy it is to mark up text to produce typeset output. All you have to do is to type the (electronic) source file—LaTeX does the rest.

## *1.1 The source file*

In the next several sections, you will be introduced to the most important commands for typesetting text by working through examples.

The source file is made up of *text, math* (e.g., $\sqrt{5}$), and *instructions to LaTeX*. This is how you type the last sentence:

```
The source file is made up of \emph{text,} \emph{math} (e.g.,
$\sqrt{5}$), and \emph{instructions to \LaTeX.}
```

In that sentence,

```
The source file is made up of \emph{text,} \emph{math} (e.g.,
```

is text,

```
$\sqrt{5}$
```

is math, and

```
\emph{instructions to \LaTeX}
```

is an instruction (a command with an argument). Commands, as a rule, start with a backslash (\) and tell LaTeX to do something special. In this case, the command `\emph` emphasizes its ***argument*** (the text between the braces). Another kind of instruction is called an ***environment***. For instance, the commands

```
\begin{flushright}
```

and

```
\end{flushright}
```

enclose a `flushright` environment—text that is typed inside this environment is right justified (lined up against the right margin) when typeset. (The `flushleft` environment creates left-justified text; the `center` environment creates text that is centered on the page.)

In practice, text, math, and instructions are usually intertwined. For example,

```
\emph{My first integral} $\int \zeta^{2}(x) \, dx$.
```

produces

*My first integral* $\int \zeta^{2}(x) \, dx$.

which is a mixture of all three. Despite this, we will discuss the three topics (typing text, typing math, and giving instructions to LaTeX) as if they were independent in order to make the discussion clearer.

We will be working with a number of sample documents, which can be typed from the examples in the text or downloaded from the Internet (see Section C.1). Create a directory on your computer called `samples` to store the sample files, and another directory called `work` where you can keep your working files. Whenever you want to use one of these documents, copy it from the `samples` directory into the `work` directory, so that the original remains unchanged. *In this book, the* `samples` *and* `work` *directories will refer to the directories you have created.*

## 1.2 The keyboard

The following keys are used to type text in a LaTeX source file:

```
a-z   A-Z   0-9
+ = * / ( ) [ ]
```

You may also use the following punctuation marks:

```
, ; . ? ! : ‘ ’ -
```

and the spacebar, the tab key (which—unlike its function in a word processor—has the same effect as the spacebar), and the Return (or Enter) key.

There is one possible problem affecting the portability of LaTeX source files: The Return (or Enter) key writes an invisible code into your source file that indicates where the line ends. Because this code is different on each of the major platforms (Macintosh, PC, and UNIX), you may have problems reading a source file that was created on a different platform. Luckily, many text editors include the ability to switch end-of-line codes; some even do so automatically.

Finally, there are thirteen special keys that are mostly used in LaTeX instructions:

```
# $ % & ~ _ ^ \ { } @ " |
```

There are commands available so that you can typeset most of these special characters (as well as composite characters, such as accented characters) if you need to use them in your document. For instance, $ is typed as `\$`, the underscore (_) is typed as `\_`, and % is typed as `\%`, whereas ä is typed as `\"{a}`. However, @ is simply typed `@`. See the tables in Appendix B for more detail.

LaTeX prohibits the use of other keys on your keyboard (unless you are using a version of LaTeX that is set up to work with non-English languages—see Section 6.4.4). Do not use the computer's modifier keys (e.g., Alt, Control, Command, Option) to produce special characters because LaTeX will either ignore or misinterpret them. When trying to typeset a source file that contains a prohibited character, LaTeX will display an error message similar to the following:

```
! Text line contains an invalid character.
l.222 completely irreducible^^?
                             ^^?
```

In this message, `l.222` means that you should look at line 222 of your source file to find the problem. You must edit this line to remove the character that LaTeX cannot understand.

## 1.3 *Your first note*

We start our discussion of how to type a note in LaTeX with a simple example. Suppose you want to use LaTeX to produce the following:

It is of some concern to me that the terminology used in multi-section math courses is not uniform.

In several sections of the course on matrix theory, the term "hamiltonian-reduced" is used. I, personally, would rather call these "hyper-simple." I invite others to comment on this problem.

Of special concern to me is the terminology in the course by Prof. Rudi Hochschwabauer. Since his field is new, there is no accepted terminology. It is imperative that we arrive at a satisfactory solution.

Create a new file in the work directory with the name note1.tex and type the following, including the spacing and linebreaks shown, but not the line numbers (or copy the note1.tex file from the samples directory; see page 2):

```
1   % Sample file: note1.tex
2   % Typeset with LaTeX format
3   \documentclass{article}
4
5   \begin{document}
6   It is of some concern to me   that
7   the terminology used in  multi-section
8    math courses is not uniform.
9
10  In several sections of the course on
11  matrix theory, the  term
12   ``hamiltonian-reduced'' is used.
13   I, personally, would rather call these ``hyper-simple.'' I
14  invite others to comment on this  problem.
15
16  Of special concern to me is the terminology in the course
17  by Prof.~Rudi Hochschwabauer.
18    Since his field is new, there is
19   no accepted
20  terminology.   It is imperative
21  that we arrive at a satisfactory solution.
22  \end{document}
```

The first two lines start with %. These lines are called *comments* and are ignored by LaTeX. The % character is very useful. For example, if you want to add some notes to your source file, and you do not want those notes to appear in the typeset version of your article, you can begin those lines with a %, and LaTeX will ignore everything on them when typesetting your source file. You can also comment out part of a line:

```
simply put, we believe % Actually it's not so simple.
```

When typesetting this, LaTeX will ignore everything on the line after the % character.

The third line of the source file specifies the *document class* (article, in our case), which controls how the document will be formatted.

The text of the note is typed within the document environment, that is, between the lines

```
\begin{document}
```

and

```
\end{document}
```

Now typeset note1.tex; you should get the typeset document displayed on page 3. As you can see from this example, LaTeX is different from a word processor. It ignores the way you input and position the text, and follows only the formatting instructions given by the markup commands. LaTeX notices when you put a space or tab in the text, but it ignores *how many* spaces or tabs have been inserted. Similarly, one or more blank lines mark the end of a paragraph.

LaTeX, by default, fully justifies the text by placing a certain size space between words—the *interword space*—and a somewhat larger space between sentences—the *intersentence space*. If you have to force an interword space, you can use the \␣ command (the ␣ symbol indicates a blank space).

The ~ (tilde) command also forces an interword space, but with a difference: It keeps words together on the same line. This command is called a "tie" or "nonbreakable space."

When LaTeX encounters a period, it has to decide whether or not that period indicates the end of a sentence. It uses the following rule: A period following a capital letter (e.g., A.) is interpreted as being part of an abbreviation or an initial and will be followed by an interword space. Any other period signifies the end of a sentence and will be followed by an intersentence space.

If this rule causes problems in your document, you can follow the period with \␣ to force an interword space, or precede the period with \@ to force an intersentence space, as in

```
In 1994, it was published in the Swedish Combin.\ J\@.
Next year \dots
```

which will print correctly:

In 1994, it was published in the Swedish Combin. J. Next year ...

Note that on lines 12 and 13, the left double quotes are typed as `` (two left single quotes) and the right double quotes are typed as '' (two right single quotes). The left single quote key is not always easy to find; it is usually hidden in the upper-left or upper-right corner of the keyboard, and shares a key with the tilde (~).

## 1.4 Lines too wide

LaTeX reads the text in the source file one line at a time; when the end of a paragraph is reached, LaTeX typesets the entire paragraph. Most of the time, there is no need for corrective action. Occasionally, however, LaTeX gets into trouble when trying to split the paragraph into typeset lines. To illustrate this situation, modify `note1.tex`: On line 12, replace "`term`" by "`strange term`"; and on line 18, delete "`Rudi `". Now save this modified file with the name `note1b.tex` in the `work` directory. (If you downloaded the sample files, you can find `note1b.tex` in the `samples` directory—see page 2.)

When you typeset `note1b.tex`, you should obtain the following:

It is of some concern to me that the terminology used in multi-section math courses is not uniform.

In several sections of the course on matrix theory, the strange term "hamiltonian-reduced" is used. I, personally, would rather call these "hyper-simple." I invite others to comment on this problem.

Of special concern to me is the terminology in the course by Prof. Hochschwabauer. Since his field is new, there is no accepted terminology. It is imperative that we arrive at a satisfactory solution.

The first line of paragraph two is about 1/4 inch too wide. The first line of paragraph three is even wider. On your monitor, LaTeX displays the following messages:

```
Overfull \hbox (15.38948pt too wide) in paragraph at lines 10--15
[]\OT1/cmr/m/n/10 In sev-eral sec-tions of the course on ma-trix
the-ory, the strange term ``hamiltonian-
 []
Overfull \hbox (23.27834pt too wide) in paragraph at lines 16--22
[]\OT1/cmr/m/n/10 Of spe-cial con-cern to me is the ter-mi-nol-ogy
in the course by Prof. Hochschwabauer.
 []
```

You will find the same message in the log file, `note1b.log` (see Section 6.4).

The message

```
Overfull \hbox (15.38948pt too wide) in paragraph at lines 10--15
```

refers to paragraph two (lines 10–15 in the source file—its location in the typeset document is not specified): The typeset version of this paragraph has a line that is 15.38948 points too wide. LaTeX uses *points* (pt) to measure distances; there are about 72 points in 1 inch (or about 28 points in 1 cm); so 15.38948 points is about this long: ⌴.

The next two lines,

```
[]\OT1/cmr/m/n/10 In sev-eral sec-tions of the course on ma-trix
the-ory, the strange term ``hamiltonian-
```

identify the source of the problem: LaTeX did not hyphenate the word
`hamiltonian-reduced`
since it only hyphenates a hyphenated word *at the hyphen.*

The second reference,

```
Overfull \hbox (23.27834pt too wide) in paragraph at lines 16--22
```

is to paragraph three (lines 16–22 of the source file). There is a problem with the word `Hochschwabauer`, which LaTeX's standard hyphenation routine cannot handle. (A German hyphenation routine would have no difficulty hyphenating `Hochschwabauer`—see Section 6.4.4.) If you encounter such a problem, you can either try to reword the sentence or insert one or more *optional hyphen* commands (`\-`), which tell LaTeX where it may hyphenate the word.

In this case, you can rewrite Hochschwabauer as `Hoch\-schwabauer` and the second hyphenation problem disappears. If you plan to use a problematic word several times, you can implement a "global solution" (i.e., an instruction taking care of this problem for the whole document) by placing the line

```
\hyphenation{Hoch-schwa-bau-er}
```

before the text (in the preamble—see Section 4.1). You can put as many words as necessary in the argument of this command; separate the words with spaces.

Sometimes a small horizontal overflow can be difficult to spot. The `draft` document class option may help: LaTeX will put a black box (or *slug*) in the margin to mark an overfull line. Changing the `\documentclass` line to

```
\documentclass[draft]{article}
```

invokes this option. A version of `note1b.tex` with this option set can be found in the `samples` directory under the name `noteslug.tex`.

## 1.5 More text features

Next, we will produce the following note:

January 12, 1999

**From the desk of George Grätzer**

February 7–21 *please* use my temporary e-mail address:

George_Gratzer@umanitoba.ca

Type the following source file and save it as note2.tex in the work directory, without line numbers (note2.tex is provided in the samples directory—see page 2):

```
1   % Sample file: note2.tex
2   % Typeset with LaTeX format
3   \documentclass{article}
4
5   \begin{document}
6   \begin{flushright}
7      \today
8   \end{flushright}
9   \textbf{From the desk of George Gr\"{a}tzer}\\[22pt]
10  February~7--21 \emph{please} use my temporary e-mail address:
11  \begin{center}
12     \texttt{George\_Gratzer@umanitoba.ca}
13  \end{center}
14  \end{document}
```

This note introduces several additional features of LaTeX:

- The \today command displays the date on which the document is being typeset.
- The use of environments to *right justify* or *center* text. (Note that we indent the contents of the environment to make the source file easier to read.)
- The use of text style commands, including the \emph command to *emphasize* text, the \textbf command to **embolden** text, and the \texttt command to produce `typewriter style` text.

  These are *commands with arguments:* In each case, the argument of the command follows the name of the command and is typed between braces, that is, between { and }. Note that command names are *case sensitive:* Typing \Textbf or \TEXTBF instead of \textbf will generate an error message.

- LaTeX commands almost always start with a backslash (\) followed by the command name, for instance, `\textbf`. The command name is terminated by the first ***non-alphabetic character*** (i.e., by any character other than a–z or A–Z).
- The use of double hyphens for number ranges (en dash): `7--21` prints as 7–21; use triple hyphens (`---`) for the "em dash" punctuation mark—such as the one in this sentence.
- Breaking a line with the `\\` command (`\newline` is another form). To create additional space between lines (as in the last note, under the line **From the desk** ... ), you can use the `\\` command and specify an appropriate amount of vertical space: `\\[22pt]` (see also Formula 20 in Section 3.1). Note that this command uses ***square brackets*** rather than braces because the argument is ***optional.*** The distance may be given in points, centimeters (cm), or inches (in).

  To force a page break, use `\newpage`.
- There are special rules for special characters (see Section 1.2), for ***accented characters,*** and for some ***European characters.*** For instance, the special character underscore (_) is typed as `\_`, and the accented character ä is typed as `\"{a}` (see the tables in Appendix B). Words with accented characters and non-English words are not properly hyphenated by LaTeX; see Section 6.4.4 for more information.

# 1.6 List environments

LaTeX provides three basic list environments for your use: `enumerate`, `itemize`, and `description`.

- The `enumerate` environment produces numbered lists:

> This space has the following properties:
>
> 1. Grade 2 Cantor
> 2. Half-smooth Hausdorff
>
> We can apply the Main Theorem ...

is typed as

```
This space has the following properties:
\begin{enumerate}
   \item Grade 2 Cantor
   \item Half-smooth Hausdorff
\end{enumerate}
We can apply the Main Theorem \ldots
```

Each item is introduced with an `\item` command. This construct can be used in theorems and definitions to list conditions and conclusions.

- You will find many examples of the `itemize` environment in this book, in which each item is marked by a bullet (for instance, this text is inside an `itemize` environment; pages xv and 8 provide further examples). The following example:

  In this lecture, we set out to accomplish a variety of goals:

  - To introduce the concept of smooth functions
  - To show their usefulness in the differentiation of Howard-type functions
  - To point out the efficacy of using smooth functions in advanced calculus courses

  is typed as

```
In this lecture, we set out to accomplish a variety of goals:
\begin{itemize}
   \item To introduce the concept of smooth functions
   \item To show their usefulness in the differentiation
      of Howard-type functions
   \item To point out the efficacy of using smooth functions
      in advanced calculus courses
\end{itemize}
```

  Because this list is an `itemize` environment within an `itemize` environment, each bullet is replaced by an en dash.

- In the `description` environment, the optional argument to the `\item` command is the term you are describing:

  **J. Perelman** the first to introduce smooth functions

  is typed as

```
\begin{description}
  \item[J. Perelman] the first to introduce smooth functions
\end{description}
```

CHAPTER

# 2 Typing math

Now we will start mixing text with mathematical formulas.

## 2.1 A note with mathematical formulas

There are three additional keys that you will need to type mathematical formulas: |, <, and >. ( | is the shifted \ key on many keyboards. You can also use these characters in text mode with the commands in Section B.3.)

You will begin typesetting math with the following note:

In first-year calculus, we define intervals such as $(u, v)$ and $(u, \infty)$. Such an interval is a *neighborhood* of $a$ if $a$ is in the interval. Students should realize that $\infty$ is only a symbol, not a number. This is important since we soon introduce concepts such as $\lim_{x \to \infty} f(x)$.

When we introduce the derivative,

$$\lim_{x \to a} \frac{f(x) - f(a)}{x - a},$$

we assume that the function is defined and continuous in a neighborhood of $a$.

To create the source file for this mixed math and text note, create a new document with your text editor. Name it `math.tex`, save it in your `work` directory, and type in the following source file—without the line numbers (or simply copy `math.tex` from the `samples` directory, see page 2):

```
1   % Sample file: math.tex
2   % Typeset with LaTeX format
3   \documentclass{article}
4
5   \begin{document}
6   In first-year calculus, we   define intervals  such as
7   $(u, v)$ and $(u, \infty)$.  Such an interval is a
8   \emph{neighborhood} of  $a$
9   if  $a$ is in the interval.  Students should
10  realize that  $\infty$ is only a
11  symbol, not a number.  This is important since
12  we soon introduce concepts
13   such as $\lim_{x \to \infty} f(x)$.
14
15  When we introduce the derivative,
16  \[
17     \lim_{x \to a} \frac{f(x) - f(a)}{x - a},
18  \]
19  we assume that the function is defined and continuous
20  in a neighborhood of  $a$.
21  \end{document}
```

This note introduces several basic concepts for typesetting math in LaTeX:

- There are two kinds of math formulas and environments:
  1. *Inline math environments* open and close with `$` (as seen in this book) or open with `\(` and close with `\)`.
  2. *Displayed math environments* open with `\[` and close with `\]`.
- Within math environments, LaTeX uses its own spacing rules and completely ignores the number of white spaces typed with two exceptions:
  1. Spaces that delimit commands (e.g., in `$\infty a$`, the space is not ignored; in fact, `$\inftya$` is an error)
  2. Spaces in the arguments of commands that temporarily revert to text mode (`\mbox` is such a command; see Section 2.3)

  The white space added when typing math is important only for the readability of the source file. To summarize:

**Rule** ■ Spacing in text and math
In text mode, many spaces equal one space, whereas in math mode, spaces are ignored (unless they terminate a command).

---

To adjust the spacing in the typeset article, use the spacing commands listed in Section A.6.

- The same formula may be typeset differently depending on whether it is inline or displayed. The expression $x \to a$ is set as a ***subscript*** to lim in the inline formula $\lim_{x \to a} f(x)$, typed as `$\lim_{x \to a} f(x)$`, but it is automatically set *below* lim in the displayed version,

$$\lim_{x \to a} f(x)$$

typed as

```
\[
   \lim_{x \to a} f(x)
\]
```

- Math symbols are invoked by commands inside a math formula or environment. For example, the command for $\infty$ is `\infty`, and the command for $\to$ is `\to`. The math symbols are organized into tables in Appendix A.

  To access all LaTeX symbols, use the latexsym package. Begin your article with

```
\documentclass{article}
\usepackage{latexsym}
```

Ⓐ Many of the symbols listed in Appendix A require the amssymb package. To use all of the LaTeX and AMS symbols, load both packages:

```
\usepackage{amssymb,latexsym}
```

- Some commands (e.g, `\sqrt`) need ***arguments*** enclosed in braces (`{` and `}`). To typeset $\sqrt{5}$, type `$\sqrt{5}$`, where `\sqrt` is the command and `5` is the argument. Some commands need more than one argument: To get

$$\frac{3+x}{5}$$

type

```
\[
   \frac{3+x}{5}
\]
```

`\frac` is the command; `3+x` and `5` are the arguments.

## 2.2 Errors in math

There can be many mistakes, even in such a simple note. To help familiarize yourself with some of the most commonly encountered LaTeX errors and their causes, we will deliberately introduce mistakes into `math.tex`. The version of `math.tex` with mistakes is `mathb.tex`. By inserting and deleting `%` signs, you will make the mistakes visible to LaTeX one at a time. (Remember that lines starting with `%` are ignored by LaTeX.) Type the following source file, and save it under the name `mathb.tex` in the `work` directory (or copy the file `mathb.tex` from the `samples` directory—see page 2). As usual, do not type the line numbers! They are shown here to help you with the exercises.

```
 1   % Sample file: mathb.tex
 2   % Typeset with LaTeX format
 3   \documentclass{article}
 4
 5   \begin{document}
 6   In first-year calculus, we   define intervals  such as
 7   %$(u, v)$ and $(u, \infty)$.  Such an interval is a
 8    $(u, v)$ and  (u, \infty)$.  Such an interval is a
 9   \emph{neighborhood} of $a$
10   if $a$ is in the interval.  Students should
11   realize that  $\infty$ is only a
12   symbol, not a number.  This is important since
13   we soon introduce concepts
14    such as $\lim_{x \to \infty} f(x)$.
15   %such as $\lim_{x \to \infty  f(x)$.
16
17   When we introduce the derivative
18  \[
19       \lim_{x \to a} \frac{f(x) - f(a)}{x - a}
20     %\lim_{x \to a} \frac{f(x) - f(a)  x - a}
21   \]
22   we assume that the function is defined and continuous
23   in a neighborhood of  $a$.
24   \end{document}
```

**Exercise 1** Note that on line 8, the third `$` is missing. When typesetting the `mathb.tex` file, LaTeX generates the following error message:

```
! Missing $ inserted.
<inserted text>
                $
```

```
l.8 ..., v)$ and    (u, \infty
                                    )$.  Such an interval is a
?
```

Since the `$` was omitted, LaTeX reads `(u, \infty)` as text; but the `\infty` command instructs LaTeX to typeset a math symbol, which can only be done in math mode. So LaTeX offers to put a `$` in front of `\infty`. LaTeX attempts a cure, but in this example it comes too late, because math mode *should* start just before `(u`.

Whenever you see the `?` prompt, you may press Return to ignore the error and continue typesetting the document (see Section 6.4.1 for other options).

**Exercise 2** Delete the `%` at the beginning of line 7 and insert a `%` at the beginning of line 8 (this eliminates the previous error); then delete the `%` at the beginning of line 15 and insert a `%` at the beginning of line 14, introducing a new error (the closing brace of the subscript is missing). Now typeset the note. You will get the error message

```
! Missing } inserted.
<inserted text>
    }
l.15 ...im_{x \to \infty f(x)$

?
```

LaTeX is telling you that a closing brace (`}`) is missing, but it is not sure where the brace should be. LaTeX noticed that the subscript started with `{`, but reached the end of the math formula before finding the matching `}`. You must look in the formula for a `{` that is not closed, and close it with a `}`.

**Exercise 3** Now delete the `%` at the beginning of line 14, and insert a `%` at the beginning of line 15, removing the previous error. Delete the `%` at the beginning of line 20 and insert a `%` at the beginning of line 19, introducing the final error (omitting the closing brace of the first argument and the opening brace of the second argument of `\frac`). Save and typeset the file. You will get the error message

```
! LaTeX Error: Bad math environment delimiter.

l.21 \]
```

This error message says that LaTeX believes that there is a bad math environment delimiter on line 21 of your source file, specifically, the `\]`. When we look at the source file, we can see that this delimiter is correct, which means that the problem must lie in the displayed formula, which is the case: LaTeX was trying to typeset

```
\lim_{x \to a} \frac{f(x) - f(a)  x - a}
```

but `\frac` requires *two* arguments. LaTeX found `f(x) - f(a) x - a` as the first argument, then found `\]`, closing the displayed math environment before a second argument was found.

See Section 6.1 for more information about finding and fixing problems in your LaTeX source files.

## 2.3 Building blocks of a formula

A formula is built from a large collection of components:

- Arithmetic
  - Subscripts and superscripts
- Binomial coefficients
- Congruences
- Delimiters
- Ellipses
- Integrals
- Math accents
- Matrices
- Operators
  - Large operators
- Roots
- Text

Tthis section will describe each of these groups, and provide examples illustrating their use.

Some of the commands in the following examples are defined in the amsmath package; in other words, to typeset these examples with the LaTeX `article` document class, your file must begin with Ⓐ

```
\documentclass{article}
\usepackage{amssymb,latexsym,amsmath}
```

But recall my recommendation from page xviii: You may begin your articles with

```
\documentclass{amsart}
\usepackage{amssymb,latexsym}
```

and ignore all of the discussions about packages and fonts. The amsmath package is automatically loaded by the `amsart` document class, so you do not need to include the line

```
\usepackage{amsmath}
```

**Arithmetic** The *arithmetic operations* $a + b$, $a - b$, $-a$, $a/b$, and $ab$ are typed as you might expect:

```
$a + b$, $a - b$, $-a$, $a / b$, $a b$
```

There are also two other forms of multiplication and one more of division: $a \cdot b$, $a \times b$, and $a \div b$. They are typed as follows:

```
$a \cdot b$,  $a \times b$,  $a \div b$
```

Displayed fractions, such as

$$\frac{1 + 2x}{x + y + xy}$$

are typed with `\frac`:

```
\[
  \frac{1 + 2x}{x + y + xy}
\]
```

The `\frac` command is seldom used inline because it can change the interline spacing of the paragraph; see the comment on page 41 about double accents for another example of this problem.

**Subscripts and superscripts** *Subscripts* are typed with _ (underscore) and *superscripts* with ^ (caret) and enclosed in braces, that is, typed between { and }. To get $a_1$, type the following characters:

| | |
|---|---|
| Enter inline math mode: | `$` |
| Type the letter a: | `a` |
| Subscript command: | `_` |
| Bracket the subscripted 1: | `{1}` |
| Exit inline math mode: | `$` |

that is, type `$a_{1}$`. Omitting the braces in this example causes no harm, but to get $a_{10}$, you *must* type `$a_{10}$`, because `$a_10$` is typeset as $a_1 0$. Further examples: $a_{i_1}$, $a^2$, and $a^{i_1}$ are typed as

```
$a_{i_{1}}$, $a^{2}$, $a^{i_{1}}$
```

**Binomial coefficients** For *binomial coefficients,* LaTeX offers the `\choose` command. For example, the inline version, ${a \choose b+c}$, is typed as

```
$a \choose {b + c}$
```

whereas the displayed version,

$${\frac{n^2-1}{2} \choose n + 1}$$

is typed as

```
\[
  \frac{n^{2} - 1}{2} \choose {n + 1}
\]
```

The amsmath package provides the \binom command for binomial coefficients. The examples shown above can be typed as Ⓐ

```
$\binom{a}{b + c}$
```

and

```
\[
   \binom{ \frac{n^{2} - 1}{2} }{n + 1}
\]
```

**Congruences** The two most important forms are:

| | | |
|---|---|---|
| $a \equiv v \pmod{\theta}$ | typed as | `$a \equiv v \pmod{\theta}$` |
| $a \equiv v \pod{\theta}$ | typed as | `$a \equiv v \pod{\theta}$` |

The \pod command requires the amsmath package. Ⓐ

The \pmod command behaves differently when the amsmath package is used: The inline version of the expression will have different spacing than the displayed version. The first of the two congruences shown above is the inline AMS version. The following is the displayed AMS version:

$$a \equiv v \pmod{\theta}$$

**Delimiters** Parenthesis-like symbols that expand vertically to enclose a formula.

$$\left( \frac{1 + x}{2 + y^{2}} \right)^{2}$$

is typed as

```
\[
   \left( \frac{1 + x}{2 + y^{2}} \right)^{2}
\]
```

The \left( and \right) commands tell LaTeX to size the parentheses correctly (relative to the height of the formula inside the parentheses). In a multiline formula, the corresponding \left and \right commands *must be in the same line of the formula* (that is, they cannot be separated by a linebreak command). See Section A.7 for a complete list of delimiters.

Two additional examples,

$$\left\| A^{2} \right\|, \quad \left| \frac{a + b}{2} \right|$$

would be typed as

```
\[
 \left\| A^{2} \right\|, \quad \left| \frac{a + b}{2} \right|
\]
```

where `\quad` is a spacing command (see Section A.6).

**Ellipses** The *ellipsis* ( ... ) in text is provided by the `\ldots` command:

A ... Z is typed as `A \ldots Z`

In formulas, the ellipsis can be printed either as *low* (or *on-the-line*) *dots* with the `\ldots` command:

$F(x_{1}, x_{2}, \ldots, x_{n})$ is typed as `$F(x_{1}, x_{2}, \ldots, x_{n})$`

or as centered dots with the `\cdots` command:

$x_{1} + x_{2} + \cdots + x_{n}$ is typed as `$x_{1} + x_{2} + \cdots + x_{n}$`

Ⓐ If you use the amsmath package, the command `\dots` will print the correct ellipsis (with the correct spacing) in most cases; if it does not, specify the proper type with the appropriate command: `\dotsc` for an ellipsis used with a comma, `\dotsb` for an ellipsis used with a binary operation or relation, or `\dotsm` for an ellipsis used with multiplication symbols.

**Integrals** The command for an *integral* is `\int`; the lower-limit is a subscript and the upper-limit is a superscript. For example, $\int_{0}^{\pi} \sin x \, dx = 2$ is typed as

```
$\int_{0}^{\pi} \sin x \, dx = 2$
```

`\,` is a spacing command (see Section A.6).

**Math accents** The four most frequently used *math accents* are:

$\bar{a}$, typed as `$\bar{a}$`
$\hat{a}$, typed as `$\hat{a}$`
$\tilde{a}$, typed as `$\tilde{a}$`
$\vec{a}$, typed as `$\vec{a}$`

See Section A.9.1 for a complete list.

**Matrices** LaTeX provides the `array` environment to typeset matrices.

`array` is a ***subsidiary math environment:*** It must be used inside a displayed math environment or within an `equation` environment (see Section 2.4).

For example,

$$\begin{array}{cccc} a+b+c & uv & x-y & 27\\ a+b & u+v & z & 134 \end{array}$$

is typed as

```
\[
   \begin{array}{cccc}
      a + b + c & uv    & x - y & 27\\
      a + b     & u + v & z     & 134
   \end{array}
\]
```

The required argument consists of a character `l`, `r`, or `c` (meaning left, right, or center alignment) for each column. All the columns in this example are centered, so the argument is `cccc`.

The amsmath package provides you with a `matrix` subsidiary math environment; using this environment, the previous example is typed as follows: Ⓐ

```
\[
   \begin{matrix}
      a + b + c & uv    & x - y & 27\\
      a + b     & u + v & z     & 134
   \end{matrix}
\]
```

Both environments separate adjacent matrix elements within a row with ampersands (`&`); rows are separated by the linebreak (`\\`) command. No linebreak is needed on the last row.

The basic form of the AMS `matrix` environment does not include delimiters. Several additional subsidiary math environments do, including

- `pmatrix` (with parentheses)
- `bmatrix` (with brackets) Ⓐ
- `vmatrix` (with vertical lines)
- `Vmatrix` (with double vertical lines)
- `Bmatrix` (with braces)

For example,

$$\mathbf{A} = \begin{pmatrix} a+b+c & uv \\ a+b & u+v \end{pmatrix} \begin{vmatrix} 30 & 7 \\ 3 & 17 \end{vmatrix}$$

is typed as follows:

```
\[
   \mathbf{A} =
   \begin{pmatrix}
      a + b + c & uv\\
      a + b & u + v
   \end{pmatrix}
   \begin{vmatrix}
      30 & 7\\
      3 & 17
   \end{vmatrix}
\]
```

As you can see, `pmatrix` typesets as a `matrix` between a pair of `\left(` and `\right)` commands, and `vmatrix` typesets as a `matrix` between a pair of `\left|` and `\right|` commands.

**Operators** To typeset the sine function, $\sin x$, type: `$\sin x$`.

Note that `$sin x$` is typeset as $sinx$, because LaTeX interprets this expression as the product of four variables.

LaTeX calls `\sin` an *operator*. Section A.8 lists many more operators—some are like `\sin`, but others produce a more complex display. For example,

```
\[
   \lim_{x \to 0} f(x) = 0
\]
```

typesets as

$$\lim_{x \to 0} f(x) = 0$$

Ⓐ What if the operator you need is not listed—for example, the operator lfp for least fixed point? The simplest solution is to use the amsmath package, and put the declaration

```
\DeclareMathOperator{\lfp}{lfp}
```

in the source file's preamble (see Section 4.1). Then you can type

```
\[
   \lfp (X - \{ 5 \}) = 3
\]
```

which typesets as

$$\mathrm{lfp}(X - \{5\}) = 3$$

**Large operators** The command for *sum* is \sum and for *product* is \prod. The following examples,

$$\sum_{i=1}^{n} x_{i}^{2}\qquad\prod_{i=1}^{n} x_{i}^{2}$$

are typed as

```
\[
  \sum_{i=1}^{n} x_{i}^{2}\qquad\prod_{i=1}^{n} x_{i}^{2}
\]
```

\qquad is a spacing command (see Section A.6) used to separate the two formulas.

Sums and products are examples of *large operators;* all of them are listed in Section A.8.1. They appear in a different style (and size) when used in an inline formula: $\sum_{i=1}^{n} x_{i}^{2}\quad\prod_{i=1}^{n} x_{i}^{2}$.

**Roots** \sqrt produces the *square root* of its argument; for instance, $\sqrt{a + 2b}$ is typed as

```
$\sqrt{a + 2b}$
```

The *n-th root,* $\sqrt[n]{5}$, requires the use of an *optional argument,* which is specified using brackets ( [ ] ):

```
$\sqrt[n]{5}$
```

**Text** You can include *text* in a formula with an \mbox command. For instance,

$$a = b, \mbox{\qquad by assumption}$$

is typed as

```
\[
   a = b, \mbox{\qquad by assumption}
\]
```

Note the spacing command \qquad (equivalent to \quad\quad) in the argument of \mbox. You could also have typed

```
\[
   a = b, \qquad \mbox{by assumption}
\]
```

because the \qquad command also works in math mode (see Section A.6).

Ⓐ If you use the amsmath package, the \text command is available as a replacement for the \mbox command. It works just like the \mbox command except that it automatically changes the size of its argument when necessary, as in $a^{\text{power}}$, typed as

```
$a^{\text{power}}$
```

## 2.4 Typing equations

The equation environment creates a displayed math formula and automatically generates an equation number. The equation

$$\int_{0}^{\pi} \sin x \, dx = 2 \tag{1}$$

is typed as

```
\begin{equation}\label{E:firstInt}
   \int_{0}^{\pi} \sin x \, dx = 2
\end{equation}
```

The equation number generated depends on how many other equations occur before the given equation. The placement and style of the equation number is dependent on the document class and packages you have loaded. (In this book, you will find equation numbers on the left—which is the AMS default—except for the sample LaTeX article on pages 53–54.)

To refer to this formula without having to remember a number (which may change if you edit your document), you can assign a *name* to the equation in the argument of a \label command. In this example, we call the first equation "firstInt" (first integral), and use the convention that the label of an equation starts with "E:", so that the complete \label command is

```
\label{E:firstInt}
```

The number of this formula is referenced with the \ref command. Its page is referenced using the \pageref command. For example, to get the reference "see (1)", type

```
see~(\ref{E:firstInt})
```

The amsmath package includes the `\eqref` command, which provides the reference number ***in parentheses.*** This command is "smart": Even if the equation number is referenced in boldface or italic text, the reference will be typeset upright and in roman type. Ⓐ

Note the use of the tie (~) to ensure that the equation number is on the same line as the word "see". You should always use ties to connect a `\ref` command to the name of its part (e.g., equation, page, section, chapter).

The main advantage of this cross-referencing system is that when you rearrange, add, or delete equations, LaTeX automatically renumbers the equations and adjusts the references in the typeset document.

---

**Rule** ■ Typeset (at least) twice
For renumbering to work, you have to typeset your source file twice.

---

LaTeX stores the labels in the aux file while it typesets the source file (see Section 6.4). For each label, it stores the number of the equation and the page on which the equation appears.

The system described here is called *symbolic referencing.* The symbol for the number is the argument of the `\label` command, and that symbol can be referenced with the `\ref`, `\eqref`, or `\pageref` commands. LaTeX uses the same mechanism for all of the numberings it automatically generates: sections, subsections, subsubsections, equations, theorems, lemmas, and bibliographic references—except that for bibliographic references, LaTeX uses the `\bibitem` command to define a bibliographic item, and the `\cite` command to cite a bibliographic item (see Section 4.4.4). Ⓐ

With the amsmath package, equations can also be *tagged* by attaching a name to the formula with the `\tag` command. The tag replaces the equation number. Ⓐ

For example,

$$\int_{0}^{\pi} \sin x \, dx = 2 \tag{Int}$$

is typed as

```
\begin{equation}
   \int_{0}^{\pi} \sin x \, dx = 2 \tag{Int}
\end{equation}
```

Tags (of the type discussed here) are ***absolute:*** This equation can ***always*** be referred to as (Int). Equation numbers, on the other hand, are ***relative:*** They may change when equations are added, deleted, or rearranged.

## 2.5 AMS *aligned formulas*

LaTeX uses the eqnarray environment to typeset multiline formulas. This environment is quite limited and somewhat complicated to use. Use the AMS environments and avoid LaTeX's eqnarray environment.

Ⓐ We will discuss three environments: align, alignat, and cases.

### 2.5.1 *The* align *environment*

The align environment may be used to align two or more formulas or to break up a long formula.

#### *Simple alignment*

Ⓐ The basic form of the align math environment allows *simple alignment*, where *each line* is automatically numbered and the lines are aligned in a single column.
Ⓐ (The align* environment is the unnumbered version of align.)

To obtain the formulas

$$r^2 = s^2 + t^2 \tag{2}$$
$$2u + 1 = v + w^\alpha \tag{3}$$
$$x = \frac{y + z}{\sqrt{s + 2u}} \tag{4}$$

type the following, using \\ as the line separator and & as the alignment point (note that you do not need a \\ on the last line):

```
\begin{align}
   r^{2}  &= s^{2} + t^{2}                  \label{E:Pyth}\\
   2u + 1 &= v + w^{\alpha}                 \label{E:alpha}\\
   x      &= \frac{y + z}{\sqrt{s + 2u}}  \label{E:frac}
\end{align}
```

(These formulas are numbered (2), (3), and (4) because they are preceded by one numbered equation earlier in this chapter.)

The align environment can also be used to break a long formula into two (or more) parts. Since numbering both lines of such a formula would be undesirable,
Ⓐ you can prevent the numbering of the second line by using the \notag command in the second part of the formula.

For example,

$$\begin{aligned} h(x) &= \int \left( \frac{f(x) + g(x)}{1 + f^2(x)} + \frac{1 + f(x)g(x)}{\sqrt{1 - \sin x}} \right) dx \\ &= \int \frac{1 + f(x)}{1 + g(x)}\, dx - 2\tan^{-1}(x - 2) \end{aligned} \tag{5}$$

is typed as follows:

first column second column

```
f(x) &= x + yz          & g(x) &= x + y + z\\
h(x) &= xy + xz + yz    & k(x) &= (x + y)(x + z)(y + z)
```

alignment points of first column

alignment points of second column

start of second column

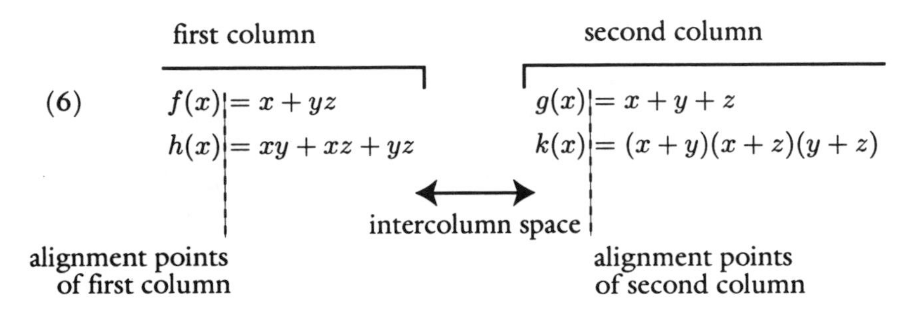


Figure 2.1: Two aligned columns: source and typeset.

```
\begin{align} \label{E:longInt}
   h(x) &= \int
    \left(
          \frac{ f(x) + g(x) }{ 1 + f^{2}(x) }
          + \frac{ 1 + f(x)g(x) }{ \sqrt{ 1 - \sin x} }
    \right) \, dx\\
        &= \int \frac{ 1 + f(x) }{ 1 + g(x) }
         \, dx - 2 \tan^{-1}(x - 2) \notag
\end{align}
```

### *Multiple alignment*

The `align` environment can also be used to align multiple columns. In the following example, there are two aligned columns: Ⓐ

$$
\begin{aligned}
f(x) &= x + yz & g(x) &= x + y + z \\
h(x) &= xy + xz + yz & k(x) &= (x + y)(x + z)(y + z)
\end{aligned} \tag{6}
$$

This example is typed as

```
\begin{align}\label{E:mm3}
   f(x) &= x + yz          & g(x) &= x + y + z\\
   h(x) &= xy + xz + yz    & k(x) &= (x + y)(x + z)(y + z)\notag
\end{align}
```

Each column is aligned and the two columns are displayed with an intercolumn space separating them.

The ampersand (&) doubles as a mark for the ***alignment point*** and as a ***column separator***, as illustrated in Figure 2.1.

### *Annotated alignment*

Ⓐ ***Annotated alignment*** allows you to align formulas and their annotations (explanatory text) separately:

$$\begin{aligned} x &= x \wedge (y \vee z) && \text{(by distributivity)}\\ &= (x \wedge y) \vee (x \wedge z) && \text{(by condition (M))}\\ &= y \vee z. \end{aligned}$$

This example is typed as

```
\begin{align*}\label{E:DoAlign}
   x &= x \wedge (y \vee z)             &&\text{(by distributivity)}\\
     &= (x \wedge y) \vee (x \wedge z) &&\text{(by condition (M))}\\
     &= y \vee z.
\end{align*}
```

In each line, in addition to the alignment point (marked by &), there is also a mark for the start of the annotation: &&.

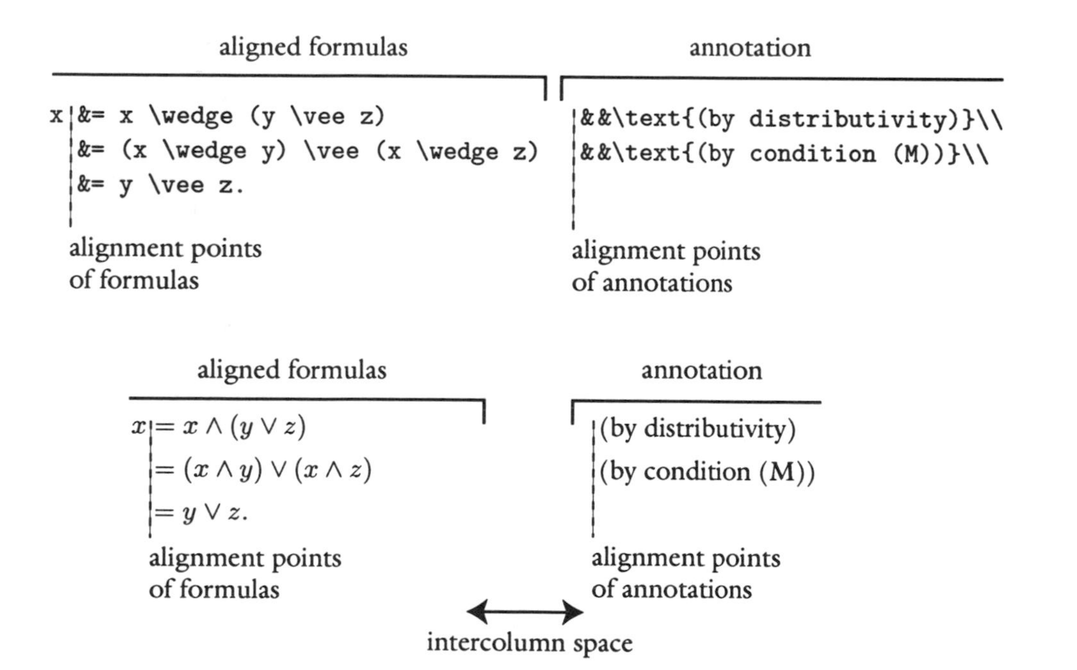


Figure 2.2: Annotated alignment: source and typeset.

### 2.5.2 *The* `alignat` *environment*

In some formulas, you may not want the intercolumn spacing that the `align` environment adds to multicolumn formulas or you may want to set your own. In such situations, use the `alignat` environment, which produces equations such as the following:

$$(A + BC)x + \qquad Cy = 0, \tag{7}$$
$$Ex + (F + G)y = 23. \tag{8}$$

typed as follows:

```
\begin{alignat}{2}
   (A + B C)&x +{} &C        &y = 0,\\
            E&x +{} &(F + G)&y = 23.
\end{alignat}
```

Here we have two columns, each aligned and pushed as close to each other as possible. (A binary operation must have something on either side; that is why the empty group (`{}`) is placed to the right of the `+`.)

There is only one new rule to learn: You must specify the number of columns in your formula as an argument. In the first line of the environment, the argument `{2}` specifies that there should be (at most) two columns.

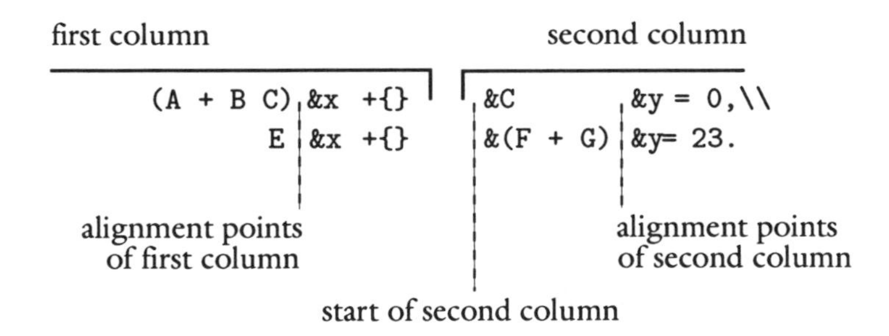


first column second column

$$(A + BC)\,x + \quad C\,y = 0, \tag{7}$$
$$E\,x + \quad (F + G)\,y = 23. \tag{8}$$

alignment points of first column

alignment points of second column

start of second column

Figure 2.3: Two columns with `alignat`: source and typeset.

How many columns will there be in the typeset formula?

**One column** if every line in the formula has one or no `&`-s

**Two columns** if the following two conditions hold:

1. Every line in the formula has *at most* three `&`-s
2. There is at least one line with two or three `&`-s

The `alignat*` environment is the unnumbered version of `alignat`.

### 2.5.3 *The* cases *environment*

Ⓐ The `cases` construct is another subsidiary math environment: It must be used inside a displayed math environment (e.g., `align`) or within an `equation` environment (see Section 2.4). Here is a typical example:

$$f(x) = \begin{cases} -x^{2}, & \text{if } x < 0; \\ \alpha + x, & \text{otherwise.} \end{cases}$$

which may be typed as follows:

```
\[
  f(x)=
  \begin{cases}
     -x^{2},         &\text{if $x < 0$;}\\
     \alpha + x,     &\text{otherwise.}
  \end{cases}
\]
```

In the `cases` environment, you type your annotation in the argument of a `\text` command and place the alignment point (`&`) before it. Lines are separated by `\\` commands.

CHAPTER

3

# Formulas and user-defined commands

The key to learning LaTeX is learning how to type formulas: We will now examine 20 sample formulas in Section 3.1 to get you started.

It should become clear from the examples in this section that some form of shorthand would make typing many of these formulas more efficient. We introduce LaTeX's shorthand system—user-defined commands—in Section 3.2.

Finally, in Section 3.3, we will build a giant formula, step-by-step.

## 3.1 Formula gallery

In this section, we present the formula gallery, a collection of formulas—some simple, some complex—that illustrate the power of LaTeX and the AMS packages. (You can find these examples in the file `gallery.tex` in the `samples` directory.) Some of the commands in these examples have not been discussed previously, but you should be able to answer most of your questions about how they work by comparing the source with the typeset results. Occasionally, we will give you a helping hand with some comments.

Many of these formulas are taken from textbooks and research articles. The last six are reproduced from the document `testart.tex`, which was distributed by the AMS some years ago.

Some of these examples require the amssymb and amsmath packages; so be sure to include the line Ⓐ

```
\usepackage{amssymb,latexsym,amsmath}
```

following the `\documentclass` line of any article using such constructs or follow my recommendation on page xviii and ignore this warning. We will point out what additional packages (if any) are required for each formula.

**Formula 1** A set-valued function:

$$x \mapsto \{\, c \in C \mid c \leq x \,\}$$

```
\[
  x \mapsto \{\, c \in C \mid c \leq x \,\}
\]
```

Note that both `|` and `\mid` are typeset as $|$. Use `|` for absolute value signs. We used `\mid` because it is a ***binary relation*** so it provides additional white space on each side. To equalize the spacing around $c \in C$ and $c \leq x$, a thin space ( `\,` ) was added inside each brace. The same technique is used in several other formulas in this section.

**Formula 2** The `\left|` and `\right|` commands are delimiters; they create vertical bars whose size adjusts to the size of the formula. The `\mathfrak` command provides access to the ***Fraktur math alphabet*** (which requires either the amsfonts or the eufrak package). Ⓐ

$$\left| \bigcup (\, I_{j} \mid j \in J \,) \right| < \mathfrak{m}$$

is typed as

```
\[
  \left| \bigcup (\, I_{j} \mid j \in J \,) \right|
   < \mathfrak{m}
\]
```

**Formula 3** Note that you have to add spacing both before and after the text fragment "for some" in the following example. The argument of `\mbox` is typeset in text mode, so the spaces are recognized.

$$A = \{\, x \in X \mid x \in X_{i}, \mbox{ for some } i \in I \,\}$$

```
\[
  A = \{\, x \in X \mid x \in X_{i},
         \mbox{ for some } i \in I \,\}
\]
```

**Formula 4** Space to show logical structure:

$$\langle a_{1}, a_{2} \rangle \leq\langle a'_{1}, a'_{2}\rangle \qquad \mbox{iff} \qquad a_{1} < a'_{1} \quad \mbox{or} \quad a_{1} = a'_{1} \mbox{ and } a_{2} \leq a'_{2}$$

```
\[
   \langle a_{1}, a_{2} \rangle \leq\langle a'_{1}, a'_{2}\rangle
    \qquad \mbox{if{f}} \qquad a_{1} < a'_{1} \quad \mbox{or}
    \quad a_{1} = a'_{1} \mbox{ and } a_{2} \leq a'_{2}
\]
```

Note that in `if{f}` (in the argument of the first `\mbox`) the second `f` is enclosed in braces to avoid the use of the ligature—the merging of the two fs.

**Formula 5** Here are some examples of Greek letters:

$$\Gamma_{u'} = \{\, \gamma \mid \gamma < 2\chi, \ B_{\alpha} \nsubseteq u', \ B_{\gamma} \subseteq u' \,\}$$

```
\[
   \Gamma_{u'} = \{\, \gamma \mid \gamma < 2\chi,
    \ B_{\alpha} \nsubseteq u', \ B_{\gamma} \subseteq u' \,\}
\]
```

Ⓐ See Section A.1.2 for a complete listing of Greek letters. The `\nsubseteq` command requires the amssymb package.

Ⓐ **Formula 6** `\mathbb` allows you to use the blackboard bold math alphabet (which only includes capital letters):

$$A = B^{2} \times \mathbb{Z}$$

```
\[
   A = B^{2} \times \mathbb{Z}
\]
```

Ⓐ Blackboard bold requires the amssymb package.

**Formula 7** The `\left(` and `\right)` delimiters (see Section 2.3) produce parentheses whose size is proportional to the height of the formula bracketed by them:

$$\left( \bigvee (\, s_{i} \mid i \in I \,) \right)^{c} = \bigwedge (\, s_{i}^{c} \mid i \in I \,)$$

```
\[
   \left( \bigvee (\, s_{i} \mid i \in I \,) \right)^{c} =
    \bigwedge (\, s_{i}^{c} \mid i \in I \,)
\]
```

Notice how the superscript is set directly above the subscript in $s_{i}^{c}$.

**Formula 8**

$$y \vee \bigvee (\, [B_{\gamma}] \mid \gamma \in \Gamma \,) \equiv z \vee \bigvee (\, [B_{\gamma}] \mid \gamma \in \Gamma \,) \pmod{ \Phi^{x} }$$

```
\[
  y \vee \bigvee (\, [B_{\gamma}] \mid \gamma
   \in \Gamma \,) \equiv z \vee \bigvee (\, [B_{\gamma}]
   \mid \gamma \in \Gamma \,) \pmod{ \Phi^{x} }
\]
```

The spacing shown in the typeset formula was created by the amsmath package. Ⓐ

**Formula 9** Use `\nolimits` to force the "limit" of a large operator to display as a subscript:

$$f(\mathbf{x}) = \bigvee\nolimits_{\!\mathfrak{m}} \left(\, \bigwedge\nolimits_{\mathfrak{m}} (\, x_{j} \mid j \in I_{i} \,) \mid i < \aleph_{\alpha} \,\right)$$

```
\[
  f(\mathbf{x}) = \bigvee\nolimits_{\!\mathfrak{m}}
   \left(\,
     \bigwedge\nolimits_{\mathfrak{m}}
     (\, x_{j} \mid j \in I_{i} \,) \mid i < \aleph_{\alpha}
   \,\right)
\]
```

The `\mathfrak` command requires either the amsfonts or the eufrak package. Notice that a negative thinspace ( `\!` ) was inserted to bring the $\mathfrak{m}$ a little closer to the large join symbol ($\bigvee$). Ⓐ

**Formula 10** The `\left.` command inserts a blank left delimiter, which is needed to balance the `\right|` command (if `\left` and `\right` commands are not balanced, you get an error message):

$$\left. \widehat{F}(x) \right|_{a}^{b} = \widehat{F}(b) - \widehat{F}(a)$$

```
\[
  \left. \widehat{F}(x) \right|_{a}^{b} =
     \widehat{F}(b) - \widehat{F}(a)
\]
```

**Formula 11**

$$u \underset{\alpha}{+} v \overset{1}{\thicksim} w \overset{2}{\thicksim} z$$

```
\[
  u \underset{\alpha}{+} v \overset{1}{\thicksim} w
    \overset{2}{\thicksim} z
\]
```

Ⓐ The \underset and \overset commands require the amsmath package. LaTeX can do a special case with the \stackrel command: Placing a symbol above a *binary relation.*

**Formula 12** In this formula, \mbox would not work properly because the overset text would be too large, so we use \text, which requires the amsmath package: Ⓐ

$$f(x) \overset{\text{def}}{=} x^{2} - 1$$

```
\[
   f(x) \overset{ \text{def} }{ = } x^{2} - 1
\]
```

**Formula 13**

$$\overbrace{a + b + \cdots + z}^{n}$$

```
\[
   \overbrace{a + b + \cdots + z}^{n}
\]
```

Ⓐ Note that if you use the amsmath package, \dots will do.

**Formula 14**

$$\begin{vmatrix} a + b + c & uv \\ a + b & c + d \end{vmatrix} = 7$$

```
\[
   \begin{vmatrix}
      a + b + c & uv\\
      a + b & c + d
   \end{vmatrix}
   = 7
\]
```

$$\begin{Vmatrix} a+b+c & uv \\ a+b & c+d \end{Vmatrix} = 7$$

```
\[
   \begin{Vmatrix}
      a + b + c & uv\\
      a + b & c + d
   \end{Vmatrix}
   = 7
\]
```

The `vmatrix` and `Vmatrix` environments require the amsmath package. Using standard LaTeX, the second matrix would be typed as Ⓐ

```
\[
   \left\|\begin{array}{cc}
      a + b + c & uv\\
      a + b & c + d
   \end{array}\right\|
   = 7
\]
```

which produces

$$\left\|\begin{array}{cc} a+b+c & uv \\ a+b & c+d \end{array}\right\| = 7$$

Note that LaTeX's spacing is different: There is more space between the vertical bars and the matrix entries. The columns are centered because of the `{cc}` argument; you could make them align flush left or flush right instead by using `l` or `r`, respectively.

**Formula 15** The `\mathbf{N}` command makes a bold **N**:

$$\sum_{j \in \mathbf{N}} b_{ij} \hat{y}_{j} = \sum_{j \in \mathbf{N}} b^{(\lambda)}_{ij} \hat{y}_{j} + (b_{ii} - \lambda_{i}) \hat{y}_{i} \hat{y}$$

```
\[
   \sum_{j \in \mathbf{N}} b_{ij} \hat{y}_{j} =
   \sum_{j \in \mathbf{N}} b^{(\lambda)}_{ij} \hat{y}_{j} +
   (b_{ii} - \lambda_{i}) \hat{y}_{i} \hat{y}
\]
```

To get a bold math symbol (in math mode), use the `\boldsymbol` command from the amsmath package: `$\boldsymbol{\alpha}$` produces $\boldsymbol{\alpha}$. Ⓐ

**Formula 16** To produce the formula

$$\biggl(\prod^n_{\, j=1} \hat x_{j} \biggr) H_{c} = \frac{1}{2} \hat{k}_{ij} \det \widehat{\mathbf{K}}(i|i)$$

try typing

```
\[
   \left( \prod^n_{\, j = 1} \hat x_{j} \right) H_{c} =
    \frac{1}{2} \hat k_{ij} \det \hat{ \mathbf{K} }(i|i)
\]
```

This gives

$$\left(\prod^n_{\, j=1} \hat x_{j} \right) H_{c} = \frac{1}{2} \hat k_{ij} \det \hat{\mathbf{K}}(i|i)$$

which isn't quite right. You can correct the overly large parentheses by using the \biggl and \biggr commands in place of \left( and \right), respectively. Adjust the small hat over $K$ by using \widehat:

```
\[
   \biggl( \prod^n_{\, j = 1} \hat x_{j} \biggr) H_{c} =
    \frac{1}{2} \hat{k}_{ij} \det \widehat{ \mathbf{K} }(i|i)
\]
```

which will give you the original formula:

$$\biggl(\prod^n_{\, j=1} \hat x_{j} \biggr) H_{c} = \frac{1}{2} \hat{k}_{ij} \det \widehat{\mathbf{K}}(i|i)$$

**Formula 17** In this formula, we use \overline{I} to get $\overline{I}$ (we could also use \bar{I}, which is set as $\bar{I}$):

$$\det \mathbf{K}(t=1, t_{1}, \ldots, t_{n}) = \sum_{I \in \mathbf{n}}(-1)^{|I|} \prod_{i \in I} t_{i} \prod_{j \in I} (D_{j} + \lambda_{j} t_{j}) \det \mathbf{A}^{(\lambda)}(\,\overline{I} | \overline{I}\,) = 0$$

```
\[
  \det \mathbf{K} (t = 1, t_{1}, \ldots, t_{n}) =
  \sum_{I \in \mathbf{n} }(-1)^{|I|}
  \prod_{i \in I} t_{i}
  \prod_{j \in I} (D_{j} + \lambda_{j} t_{j})
  \det \mathbf{A}^{(\lambda)}(\,\overline{I} | \overline{I}\,)= 0
\]
```

Note that if you use the amsmath package, \dots will do.

**Formula 18** The command \| provides the $\|$ math symbol in this formula:

$$\lim_{(v, v') \to (0, 0)} \frac{H(z + v) - H(z + v') - BH(z)(v - v')}{\| v - v' \|} = 0$$

```
\[
   \lim_{(v, v') \to (0, 0)}
    \frac{H(z + v) - H(z + v') - BH(z)(v - v')}
         {\| v - v' \|} = 0
\]
```

**Formula 19** This formula uses the calligraphic math alphabet:

$$\int_{\mathcal{D}} | \overline{\partial u} |^{2} \Phi_{0}(z) e^{\alpha |z|^2} \geq c_{4} \alpha \int_{\mathcal{D}} |u|^{2} \Phi_{0} e^{\alpha |z|^{2}} + c_{5} \delta^{-2} \int_{A} |u|^{2} \Phi_{0} e^{\alpha |z|^{2}}$$

```
\[
   \int_{\mathcal{D}} | \overline{\partial u} |^{2}
    \Phi_{0}(z) e^{\alpha |z|^2} \geq
    c_{4} \alpha \int_{\mathcal{D}} |u|^{2} \Phi_{0}
    e^{\alpha |z|^{2}} + c_{5} \delta^{-2} \int_{A}
    |u|^{2} \Phi_{0} e^{\alpha |z|^{2}}
\]
```

**Formula 20** The \hdotsfor command sets dots that span multiple columns in a matrix. The \dfrac command is the displayed variant of \frac. Ⓐ Ⓐ

$$\mathbf{A} = \begin{pmatrix} \dfrac{\varphi \cdot X_{n, 1}}{\varphi_{1} \times \varepsilon_{1}} & (x + \varepsilon_{2})^{2} & \cdots & (x + \varepsilon_{n - 1})^{n - 1} & (x + \varepsilon_{n})^{n}\\[10pt] \dfrac{\varphi \cdot X_{n, 1}}{\varphi_{2} \times \varepsilon_{1}} & \dfrac{\varphi \cdot X_{n, 2}}{\varphi_{2} \times \varepsilon_{2}} & \cdots & (x + \varepsilon_{n - 1})^{n - 1} & (x + \varepsilon_{n})^{n}\\ \hdotsfor{5}\\ \dfrac{\varphi \cdot X_{n, 1}}{\varphi_{n} \times \varepsilon_{1}} & \dfrac{\varphi \cdot X_{n, 2}}{\varphi_{n} \times \varepsilon_{2}} & \cdots & \dfrac{\varphi \cdot X_{n, n - 1}}{\varphi_{n} \times \varepsilon_{n - 1}} & \dfrac{\varphi \cdot X_{n, n}}{\varphi_{n} \times \varepsilon_{n}} \end{pmatrix} + \mathbf{I}_{n}$$

```
\[
   \mathbf{A} =
   \begin{pmatrix}
      \dfrac{\varphi \cdot X_{n, 1}}
            {\varphi_{1} \times \varepsilon_{1}}
      & (x + \varepsilon_{2})^{2} & \cdots
      & (x + \varepsilon_{n - 1})^{n - 1}
      & (x + \varepsilon_{n})^{n}\\[10pt]
```

```
      \dfrac{\varphi \cdot X_{n, 1}}
            {\varphi_{2} \times \varepsilon_{1}}
      & \dfrac{\varphi \cdot X_{n, 2}}
              {\varphi_{2} \times \varepsilon_{2}}
      & \cdots & (x + \varepsilon_{n - 1})^{n - 1}
      & (x + \varepsilon_{n})^{n}\\
      \hdotsfor{5}\\
      \dfrac{\varphi \cdot X_{n, 1}}
            {\varphi_{n} \times \varepsilon_{1}}
      & \dfrac{\varphi \cdot X_{n, 2}}
              {\varphi_{n} \times \varepsilon_{2}}
      & \cdots & \dfrac{\varphi \cdot X_{n, n - 1}}
                        {\varphi_{n} \times \varepsilon_{n - 1}}
      & \dfrac{\varphi\cdot X_{n, n}}
              {\varphi_{n} \times \varepsilon_{n}}
   \end{pmatrix}
    + \mathbf{I}_{n}
\]
```

Ⓐ This formula requires both the amsmath and the amssymb packages. In the next section, you will learn how to type it in a shorter and more readable form. Note the use of the command `\\[10pt]` (see Section 1.5); if you only use `\\`, the first and second lines of the matrix are displayed too close together.

# 3.2 User-defined commands

You can make LaTeX easier to use if you judiciously add ***user-defined commands*** (***macros***) for your particular needs.

## 3.2.1 Commands as shorthand

If you use the `\leftarrow` command a lot, you could define

```
\newcommand{\la}{\leftarrow}
```

which allows you to just type `\la` to obtain a left arrow.

Instead of `\widetilde{a}`, you could type `\wa` after defining a `\wa` command as follows (we will discuss how to generalize such commands in Section 3.2.2):

```
\newcommand{\wa}{\widetilde{a}}
```

If you use the construct $D^{[2]} \times D^{[3]}$ often, you could define

```
\newcommand{\DD}{D^{[2]} \times D^{[3]}}
```

and then type `\DD` instead of `D^{[2]} \times D^{[3]}` throughout the rest of your document.

You can also define commands as a shorthand for text. For instance, if you used the phrase "`subdirectly irreducible`" many times in your source file, you could define

```
\newcommand{\si}{subdirectly irreducible}
```

and `\si` would become shorthand for `subdirectly irreducible`.

**Rule** ■ User-defined commands

**Definition**

1. Type the `\newcommand` command.
2. In braces, type the name of the new command, ***including the backslash*** (`\`).
3. In a second set of braces, define the new command.

**Use**

4. Use new commands such as `\si` (defined above) as `\si\␣` or `\si{}` before a space or an alphabetical character, and `\si` otherwise.

To illustrate Rule 4, typeset `\si{} lattice` (alternatively: `\si\ lattice` or `{\si} lattice`) to produce the phrase "subdirectly irreducible lattice". *Do not type* `\si lattice`, which typesets as "subdirectly irreduciblelattice". (Of course, this rule applies to *all* commands.)

You should group all of the user-defined commands in your source file together in the preamble (see Section 4.1) between the `\usepackage` lines and the `\begin{document}` line. You will then be able to quickly find the definition of a command. This is particularly important if the source file is shared with co-authors or editors, who may need to modify your commands or add their own.

### 3.2.2 *Commands with arguments*

If you define (using the amsmath package)

```
\newcommand{\Ahh}{\Hat{\Hat{A}}}
```

then you can use the `\Ahh` command in place of `\Hat{\Hat{A}}`. If you use double hats on numerous characters, however, you may want a command for adding double hats to *any* character. Here is the command to do that:

```
\newcommand{\hh}[1]{\Hat{\Hat{#1}}}
```

Now, to get $\hat{\hat{A}}$, you can type `$\hh{A}$`. The form of this `\newcommand` is the same as before, except that after the name of the command (`{\hh}`), we put the ***number of arguments*** in brackets; in this example, `[1]`. This allows us to use `#1` to stand in for the argument in the definition of the command. When the command is invoked, its argument is substituted for the `#1` in the definition: Typing `$\hh{B}$` results in $\hat{\hat{B}}$, `$\hh{C}$` in $\hat{\hat{C}}$. (Notice that these examples disrupt the normal spacing between lines—you should think twice about using double accents in inline formulas!)

Formula 20 in the formula gallery (see page 38) is a good candidate for using user-defined commands. If we define the following two commands:

```
\newcommand{\quot}[2]{
   \dfrac{\varphi \cdot X_{n, #1}}
   {\varphi_{#2} \times \varepsilon_{#1}}}
\newcommand{\exn}[1]{(x+\varepsilon_{#1})^{#1}}
```

then typing

```
\[
   \quot{2}{3} \qquad \exn{n}
\]
```

results in

$$\frac{\varphi \cdot X_{n,2}}{\varphi_3 \times \varepsilon_2} \qquad (x+\varepsilon_n)^n$$

With these user-defined commands, we can now rewrite Formula 20 as follows:

```
\[
   \mathbf{A} =
   \begin{pmatrix}
     \quot{1}{1} & \exn{2} & \cdots & \exn{n - 1}&\exn{n}\\[10pt]
     \quot{1}{2} & \quot{2}{2} & \cdots & \exn{n - 1} &\exn{n}\\
     \hdotsfor{5}\\
     \quot{1}{n} & \quot{2}{n} & \cdots &
     \quot{n - 1}{n} & \quot{n}{n}
   \end{pmatrix}
   + \mathbf{I}_{n}
\]
```

Observe how much shorter this form is than the version in Formula 20; it is also much easier to read. Here, once again, is the typeset formula:

$$\mathbf{A} = \begin{pmatrix} \dfrac{\varphi \cdot X_{n,1}}{\varphi_1 \times \varepsilon_1} & (x + \varepsilon_2)^2 & \cdots & (x + \varepsilon_{n-1})^{n-1} & (x + \varepsilon_n)^n \\ \dfrac{\varphi \cdot X_{n,1}}{\varphi_2 \times \varepsilon_1} & \dfrac{\varphi \cdot X_{n,2}}{\varphi_2 \times \varepsilon_2} & \cdots & (x + \varepsilon_{n-1})^{n-1} & (x + \varepsilon_n)^n \\ \hdotsfor{5} \\ \dfrac{\varphi \cdot X_{n,1}}{\varphi_n \times \varepsilon_1} & \dfrac{\varphi \cdot X_{n,2}}{\varphi_n \times \varepsilon_2} & \cdots & \dfrac{\varphi \cdot X_{n,n-1}}{\varphi_n \times \varepsilon_{n-1}} & \dfrac{\varphi \cdot X_{n,n}}{\varphi_n \times \varepsilon_n} \end{pmatrix} + \mathbf{I}_n$$

### 3.2.3 *Redefining commands*

LaTeX checks to make sure that you do not inadvertently define a new command with the same name as an existing command. To test this, try defining

```
\newcommand{\or}{\vee}
```

When you typeset your document, LaTeX will generate the error message

```
! LaTeX Error: Command \or already defined.

1.12 \newcommand{\or}{\vee}
```

If you need to redefine an existing command, you should use the `\renewcommand` command to do so. As an example, assuming that you have already defined the `\exn` command as shown in the last section, you could use `\renewcommand` to redefine `\exn` as follows:

```
\renewcommand{\exn}[1]{\langle#1\rangle}
```

**Rule** ■ Redefining commands

Do not redefine commands unless you must! Blind redefinition is the route to madness.

## 3.3 *Building a formula step-by-step*

It is easy to build up complicated formulas from the components described in Section 2.3. Try the following formula:

$$\sum_{i=1}^{\left[\frac{n}{2}\right]} \binom{x_{i,i+1}^{i^2}}{\left[\frac{i+3}{3}\right]} \frac{\sqrt{\mu(i)^{\frac{3}{2}}(i^2-1)}}{\sqrt[3]{\rho(i)-2}+\sqrt[3]{\rho(i)-1}}$$

(We will use the AMS `\binom` command here; if you want to stick with LaTeX, you Ⓐ

can use the \choose command instead.) You should build this formula up in several steps. Create a new file in your work directory. Call it formula.tex, type in the following lines, and save it:

```
% File: formula.tex
% Typeset with LaTeX format
\documentclass{article}
\usepackage{amssymb,latexsym,amsmath}
\begin{document}
\end{document}
```

(Using standard LaTeX, the fourth line should be \usepackage{latexsym}.)

At present, the file has an empty document environment. Type each part of the formula as an inline or displayed formula within this environment so that you can typeset the document and check for errors as you go.

**Step 1** We will start with $\left[ \frac{n}{2} \right]$:

```
$\left[ \frac{n}{2} \right]$
```

Type this into formula.tex and test it by typesetting the document.

**Step 2** Now you can do the sum:

$$\sum_{i = 1}^{ \left[ \frac{n}{2} \right] }$$

For the superscript, you can copy and paste the formula created in Step 1 (without the dollar signs), so that you have the following:

```
\[
   \sum_{i = 1}^{ \left[ \frac{n}{2} \right] }
\]
```

**Step 3** Next, do the two formulas in the binomial:

$$x_{i, i + 1}^{i^{2}} \qquad \left[ \frac{i + 3}{3} \right]$$

Type them as separate formulas in formula.tex:

```
\[
   x_{i, i + 1}^{i^{2}} \qquad \left[ \frac{i + 3}{3} \right]
\]
```

**Step 4** Now it is easy to do the binomial. Piece together the following formula by copying and pasting the previous formulas (dropping the \qquad command):

```
\[
  \binom{ x_{i,i + 1}^{i^{2}} }{ \left[ \frac{i + 3}{3} \right] }
\]
```

which typesets as

$$\binom{ x_{i,i + 1}^{i^{2}} }{ \left[ \frac{i + 3}{3} \right] }$$

**Step 5** Next, type the formula under the square root, $\mu(i)^{\frac{3}{2}}(i^{2} - 1)$:

```
$\mu(i)^{ \frac{3}{2} } (i^{2} - 1)$
```

and then the square root, $\sqrt{\mu(i)^{\frac{3}{2}}(i^{2} - 1)}$:

```
$\sqrt{ \mu(i)^{ \frac{3}{2} } (i^{2} - 1) }$
```

**Step 6** The two cube roots, $\sqrt[3]{\rho(i) - 2}$ and $\sqrt[3]{\rho(i) - 1}$, are easy to type:

```
$\sqrt[3]{ \rho(i) - 2 }$  $\sqrt[3]{ \rho(i) - 1 }$
```

**Step 7** Now create the fraction,

$$\frac{ \sqrt{ \mu(i)^{ \frac{3}{2}} (i^{2} -1) } }{ \sqrt[3]{\rho(i) - 2} + \sqrt[3]{\rho(i) - 1} }$$

which is typed, copied, and pasted together as

```
\[
  \frac{ \sqrt{ \mu(i)^{ \frac{3}{2}} (i^{2} -1) } }
       { \sqrt[3]{\rho(i) - 2} + \sqrt[3]{\rho(i) - 1} }
\]
```

**Step 8** Finally, the whole formula,

$$\sum_{i=1}^{\left[ \frac{n}{2} \right]} \binom{ x_{i,i + 1}^{i^{2}} }{ \left[ \frac{i + 3}{3} \right] } \frac{ \sqrt{ \mu(i)^{ \frac{3}{2}} (i^{2} -1) } }{ \sqrt[3]{\rho(i) - 2} + \sqrt[3]{\rho(i) - 1} }$$

is formed by copying and pasting the pieces together, leaving only one pair of displayed math delimiters:

```
\[
   \sum_{i = 1}^{ \left[ \frac{n}{2} \right] }
      \binom{ x_{i, i + 1}^{i^{2}} }
            { \left[ \frac{i + 3}{3} \right] }
      \frac{ \sqrt{ \mu(i)^{ \frac{3}{2}} (i^{2} - 1) } }
           { \sqrt[3]{\rho(i) - 2} + \sqrt[3]{\rho(i) - 1} }
\]
```

Note the use of

- Spacing to help distinguish the braces (some text editors will help you balance the braces)
- Separate lines for the various pieces of the formula

It is to your advantage to keep your source file readable. LaTeX does not care how its input is formatted, and would happily accept the following:

```
\[\sum_{i=1}^{\left[\frac{n}{2}\right]}\binom{x_{i,i+1}^{i^{2}}}
{\left[\frac{i+3}{3}\right]}\frac{\sqrt{\mu(i)^{\frac{3}
{2}}(i^{2}-1)}}{\sqrt[3]{\rho(i)-2}+\sqrt[3]{\rho(i)-1}}\]
```

This haphazard style will make solving problems difficult if you make a mistake, and will also make it more difficult for your co-authors or editor to work with your source file.

CHAPTER

# 4

# The anatomy of an article

In this chapter, we will discuss the anatomy of an article using the popular LaTeX `article` document class by examining the sample article `intrart.tex`. Type it in (or copy it from the `samples` directory—see page 2) as we discuss the various parts that make up an article. We will discuss the additional requirements (and features) of an article that uses the AMS enhancements in the next chapter.

## 4.1 The source file of a LaTeX article

The *preamble* of an article is everything from the first line of the source file up to the line

```
\begin{document}
```

See Figure 4.1. The preamble contains instructions that affect the entire document. The `\documentclass` command is the *only* compulsory command in the preamble. There are other commands (such as `\usepackage`) that must be placed in the preamble if they are used, but these commands do not have to be present in every document.

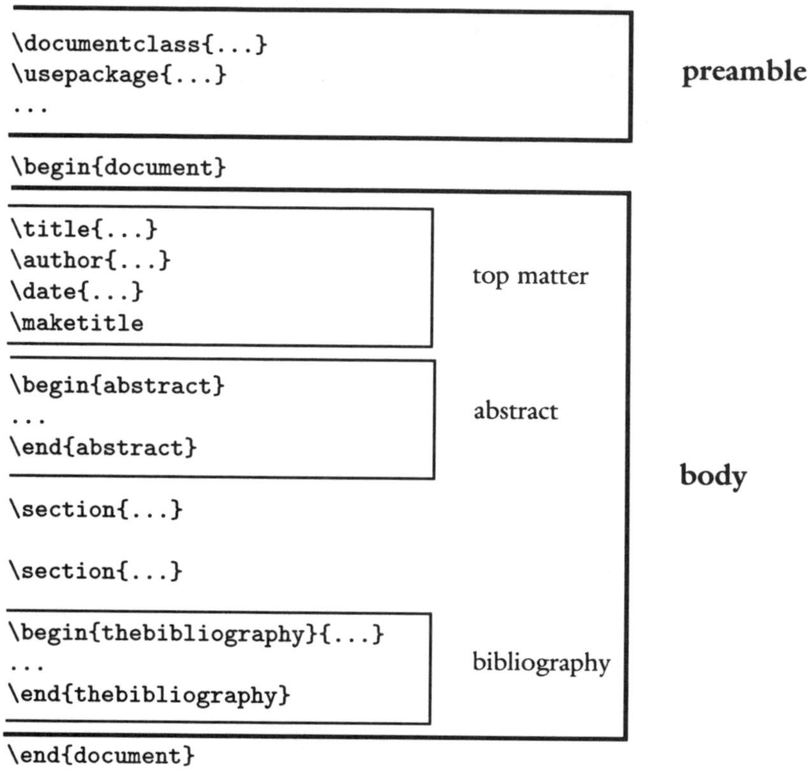

```
\documentclass{...}
\usepackage{...}
...
```

**preamble**

```
\begin{document}
```

```
\title{...}
\author{...}
\date{...}
\maketitle
```

top matter

```
\begin{abstract}
...
\end{abstract}
```

abstract

**body**

```
\section{...}

\section{...}
```

```
\begin{thebibliography}{...}
...
\end{thebibliography}
```

bibliography

```
\end{document}
```

Figure 4.1: A schematic view of an article.

Here is the preamble of the introductory sample article:

```
% Introductory sample article: intrart.tex
% Typeset with LaTeX format

\documentclass{article}
\usepackage{latexsym}
\newtheorem{theorem}{Theorem}
\newtheorem{definition}{Definition}
\newtheorem{notation}{Notation}
```

The preamble specifies the document class and then the LaTeX enhancements, or ***packages***, used by the article. It can also specify additional commands that will be used throughout the document (e.g., proclamation definitions, user-defined commands). `intrart.tex` specifies the `article` document class, and then loads the latexsym package that provides the names of some LaTeX symbols.

A ***proclamation*** is a theorem, definition, corollary, note, or other similar construct. The article `intrart.tex` defines three proclamations. One of them,

```
\newtheorem{theorem}{Theorem}
```

defines the theorem environment, which can then be used in the body of your article (see Section 4.4.3). The other two are similar. LaTeX will automatically number and format the theorems, definitions, and notations.

The article proper, called the *body*, is contained within the document environment—between the lines

```
\begin{document}
```

and

```
\end{document}
```

as illustrated in Figure 4.1. The body of the article is also split up into several parts, starting with the *top matter*, which contains the title page information. The top matter follows the line

```
\begin{document}
```

and concludes with the line

```
\maketitle
```

Here is the top matter of the introductory sample article:

```
\title{A construction of complete-simple\\
   distributive lattices}
\author{George~A. Menuhin\thanks{Research supported
   by the NSF under grant number~23466.}\\
   Computer Science Department\\
   Winnebago, Minnesota 53714\\
   menuhin@cc.uwinnebago.edu}
\date{March 15, 1999}
\maketitle
```

The body continues with an (optional) abstract, contained within an abstract environment:

```
\begin{abstract}
   In this note, we prove that there exist \emph{complete-simple
   distributive lattices,} that is, complete distributive
   lattices in which there are only two complete congruences.
\end{abstract}
```

And here is the rest of the body of the introductory sample article (with one comment in the middle), exclusive of the bibliography:

```
\section{Introduction}\label{S:intro}
In this note, we prove the following result:

\begin{theorem}
   There exists an infinite complete distributive lattice $K$
   with only the two trivial complete congruence relations.
\end{theorem}

\section{The $\Pi^{*}$ construction}\label{S:P*}
The following construction is crucial in the proof of our Theorem:

\begin{definition}\label{D:P*}
   Let $D_{i}$, for $i \in I$, be complete distributive
   lattices satisfying condition~\textup{(J)}.  Their
   $\Pi^{*}$ product is defined as follows:
   \[
      \Pi^{*} ( D_{i} \mid i \in I ) =
      \Pi ( D_{i}^{-} \mid i \in I ) + 1;
   \]
   that is, $\Pi^{*} ( D_{i} \mid i \in I )$ is
   $\Pi ( D_{i}^{-} \mid i \in I )$ with a new unit element.
\end{definition}
```

Notice that we refer to condition (J) in the definition as `\textup{(J)}`. As a result, even if the text of the definition is emphasized (as it will be in the typeset article), `(J)` will still be typeset upright, as (J), and not slanted, as *(J)*.

```
\begin{notation}
   If $i \in I$ and $d \in D_{i}^{-}$, then
   \[
     \langle \ldots, 0, \ldots, d, \ldots, 0, \ldots \rangle
   \]
   is the element of $\Pi^{*} ( D_{i} \mid i \in I )$ whose
   $i$-th component is $d$ and all the other components
   are $0$.
\end{notation}

See also Ernest~T. Moynahan~\cite{eM57a}.

Next we verify the following result:

\begin{theorem}\label{T:P*}
   Let $D_{i}$, $i \in I$, be complete distributive
```

```
   lattices satisfying condition~\textup{(J)}.  Let $\Theta$
   be a complete congruence relation on
   $\Pi^{*} ( D_{i} \mid i \in I )$.
   If there exist $i \in I$ and $d \in D_{i}$ with
   $d < 1_{i}$ such that, for all $d \leq c < 1_{i}$,
   \begin{equation}\label{E:cong1}
     \langle \ldots, d, \ldots, 0, \ldots \rangle \equiv
     \langle \ldots, c, \ldots, 0, \ldots \rangle \pmod{\Theta},
   \end{equation}
   then $\Theta = \iota$.
\end{theorem}

\emph{Proof.} Since
\begin{equation}\label{E:cong2}
   \langle \ldots, d, \ldots, 0, \ldots \rangle \equiv
   \langle \ldots, c, \ldots, 0, \ldots \rangle \pmod{\Theta},
\end{equation}
and $\Theta$ is a complete congruence relation, it follows
from condition~(J) that
\begin{equation}\label{E:cong}
  \langle \ldots, d, \ldots, 0, \ldots \rangle \equiv
  \bigvee ( \langle \ldots, c, \ldots, 0, \ldots \rangle
   \mid d \leq c < 1 ) \pmod{\Theta}.
\end{equation}
Let $j \in I$, $j \neq i$, and let $a \in D_{j}^{-}$.
Meeting both sides of the congruence (\ref{E:cong2}) with
$\langle \ldots, a, \ldots, 0, \ldots \rangle$, we obtain that
\begin{equation}\label{E:comp}
   0 = \langle \ldots, a, \ldots, 0, \ldots \rangle \pmod{\Theta},
\end{equation}
Using the completeness of $\Theta$ and (\ref{E:comp}),
we get:
\[
   0 \equiv \bigvee ( \langle \ldots, a, \ldots, 0, \ldots
    \rangle \mid a \in D_{j}^{-} ) = 1 \pmod{\Theta},
\]
hence $\Theta = \iota$.
```

At the end of the body, the *bibliographic entries* are typed between the lines

```
\begin{thebibliography}{9}
```

and

```
\end{thebibliography}
```

There are fewer than 10 references in this article, so we tell LaTeX to make room for single-digit numbering by providing the argument "9" to the thebibliography environment; you should use "99" if the number of references is between 10 and 99. The bibliography will be titled "References".

Here is the bibliography from intrart.tex:

```
\begin{thebibliography}{9}
   \bibitem{sF90}
     Soo-Key Foo,
     \emph{Lattice Constructions,} Ph.D. thesis,
     University of Winnebago, Winnebago, MN, December, 1990.
   \bibitem{gM68}
     George~A. Menuhin,
     \emph{Universal Algebra,}
     D.~van Nostrand, Princeton-Toronto-London-Melbourne, 1968.
   \bibitem{eM57}
     Ernest~T. Moynahan,
     \emph{On a problem of M.H. Stone,}
     Acta Math. Acad. Sci. Hungar. \textbf{8} (1957), 455--460.
   \bibitem{eM57a}
     Ernest~T. Moynahan,
     \emph{Ideals and congruence relations in lattices.~II,}
     Magyar Tud. Akad. Mat. Fiz. Oszt. K\"{o}zl. \textbf{9}
     (1957), 417--434.
\end{thebibliography}
```

The body (and the article) ends when the document environment is closed with

```
\end{document}
```

## 4.2 A LaTeX article typeset

The typeset version of the introductory sample article appears on the following two pages. Note that the equation numbers are on the right, which is the default in LaTeX's article document class. Elsewhere in this book you will find equation numbers on the left, which is the AMS default.

# A construction of complete-simple distributive lattices

George A. Menuhin*
Computer Science Department
Winnebago, Minnesota 23714
menuhin@cc.uwinnebago.edu

March 15, 1999

**Abstract**

In this note, we prove that there exist *complete-simple distributive lattices*, that is, complete distributive lattices in which there are only two complete congruences.

## 1 Introduction

In this note, we prove the following result:

**Theorem 1** *There exists an infinite complete distributive lattice $K$ with only the two trivial complete congruence relations.*

## 2 The $\Pi^*$ construction

The following construction is crucial in the proof of our Theorem:

**Definition 1** *Let $D_i$, for $i \in I$, be complete distributive lattices satisfying condition* (J). *Their $\Pi^*$ product is defined as follows:*

$$\Pi^*(D_i \mid i \in I) = \Pi(D_i^- \mid i \in I) + 1;$$

*that is, $\Pi^*(D_i \mid i \in I)$ is $\Pi(D_i^- \mid i \in I)$ with a new unit element.*

**Notation 1** *If $i \in I$ and $d \in D_i^-$, then*

$$\langle \ldots, 0, \ldots, d, \ldots, 0, \ldots \rangle$$

*is the element of $\Pi^*(D_i \mid i \in I)$ whose $i$-th component is $d$ and all the other components are* 0.

*Research supported by the NSF under grant number 23466.

See also Ernest T. Moynahan [4].

Next we verify the following result:

**Theorem 2** *Let $D_i$, $i \in I$, be complete distributive lattices satisfying condition* (J). *Let $\Theta$ be a complete congruence relation on $\Pi^*(D_i \mid i \in I)$. If there exist $i \in I$ and $d \in D_i$ with $d < 1_i$ such that, for all $d \leq c < 1_i$,*

$$\langle \ldots, d, \ldots, 0, \ldots \rangle \equiv \langle \ldots, c, \ldots, 0, \ldots \rangle \pmod{\Theta}, \tag{1}$$

*then $\Theta = \iota$.*

*Proof.* Since

$$\langle \ldots, d, \ldots, 0, \ldots \rangle \equiv \langle \ldots, c, \ldots, 0, \ldots \rangle \pmod{\Theta}, \tag{2}$$

and $\Theta$ is a complete congruence relation, it follows from condition (J) that

$$\langle \ldots, d, \ldots, 0, \ldots \rangle \equiv \bigvee(\langle \ldots, c, \ldots, 0, \ldots \rangle \mid d \leq c < 1) \pmod{\Theta}. \tag{3}$$

Let $j \in I$, $j \neq i$, and let $a \in D_j^-$. Meeting both sides of the congruence (2) with $\langle \ldots, a, \ldots, 0, \ldots \rangle$, we obtain that

$$0 = \langle \ldots, a, \ldots, 0, \ldots \rangle \pmod{\Theta}, \tag{4}$$

Using the completeness of $\Theta$ and (4), we get:

$$0 \equiv \bigvee(\langle \ldots, a, \ldots, 0, \ldots \rangle \mid a \in D_j^-) = 1 \pmod{\Theta},$$

hence $\Theta = \iota$.

# References


[1] Soo-Key Foo, *Lattice Constructions,* Ph.D. thesis, University of Winnebago, Winnebago, MN, December, 1990.

[2] George A. Menuhin, *Universal Algebra,* D. van Nostrand, Princeton-Toronto-London-Melbourne, 1968.

[3] Ernest T. Moynahan, *On a problem of M.H. Stone,* Acta Math. Acad. Sci. Hungar. **8** (1957), 455–460.

[4] Ernest T. Moynahan, *Ideals and congruence relations in lattices. II,* Magyar Tud. Akad. Mat. Fiz. Oszt. Közl. **9** (1957), 417–434.

## 4.3 LaTeX article templates

Before you start writing your first article, create two article templates using the article document class of LaTeX:

- article.tpl for articles with one author
- article2.tpl for articles with two authors

You will find copies of these templates in the samples directory (see page 2). Start by copying them to your work directory, or type them in from the following listings:

```
% Sample file: article.tpl
% Typeset with LaTeX format

\documentclass{article}
\usepackage{amssymb,latexsym,amsmath}

\newtheorem{theorem}{Theorem}
\newtheorem{lemma}{Lemma}
\newtheorem{proposition}{Proposition}
\newtheorem{definition}{Definition}
\newtheorem{corollary}{Corollary}
\newtheorem{notation}{Notation}

\begin{document}
\title{titleline1\\
       titleline2}
\author{name\thanks{support}\\
   addressline1\\
   addressline2\\
   addressline3}
\date{date}
\maketitle

\begin{abstract}
   abstract text
\end{abstract}

\begin{thebibliography}{99}
   bibliographic entries
\end{thebibliography}
\end{document}
```

article2.tpl is identical to article.tpl except for the argument of the \author command, which has been modified to accommodate two authors:

```
\author{name1\thanks{support1}\\
   address1line1\\
   address1line2\\
   address1line3
   \and
   name2\thanks{support2}\\
   address2line1\\
   address2line2\\
   address2line3}
```

Note the use of the \and command, which separates the two authors.

Once the template files have been copied into your work directory, you can customize them by putting your own information into the arguments of the top-matter commands. You may also want to save the modified templates in another directory, with more meaningful names (see ggart.tpl and ggart2.tpl).

The top matter of my personalized template file looks like this:

```
\title{titleline1\\
       titleline2}
\author{G. Gr\"{a}tzer\thanks{Research supported by the
                             NSERC of Canada.}\\
   University of Manitoba\\
   Department of Mathematics\\
   Winnipeg, MB R3T 2N2\\
   Canada}
\date{date}
```

Notice that the \title lines (and the \date command) were not edited because they change from article to article. For the same reason, the information concerning the second author was left unchanged in ggart2.tpl.

## 4.4 *Your first article*

Your first article will be typeset using the article document class. To start, open the personalized article template that you created in Section 4.3, and save it under the name of your first article. The name should be *one word* (no spaces, no special characters) and end with .tex.

### 4.4.1 *Editing the top matter*

Edit the top matter to contain the relevant information (e.g., title and date) for your article. Here are some simple rules to follow:

**Rule** ■ Top matter for the `article` document class

1. If necessary, separate individual lines of the title with `\\`. Do not put a `\\` at the end of the last line.
2. `\thanks` places a footnote at the bottom of the first page. If it is not needed, delete it.
3. Separate the lines of the address with `\\`. Do not put a `\\` at the end of the last line.
4. Multiple authors are separated by `\and`. There is only one `\author` command, and it contains all the information (name, address, support) about all the authors.
5. If there is no `\date` command, LaTeX will insert the date on which you are typesetting the file (`\date{\today}` will produce the same results). If you do not want *any* date to appear, type `\date{}`. For a specific date, such as "February 21, 1999," type `\date{February 21, 1999}`.
6. The `\title` command is the only compulsory command. The others are optional.

### 4.4.2 *Sectioning the body*

An article, as a rule, is divided into sections. To start the section entitled "Introduction," type

```
\section{Introduction}\label{S:intro}
```

`Introduction` is the title of the section; `S:intro` is its label, using the convention that "`S:`" starts the label for a section. The section's number is automatically assigned by LaTeX, and you can refer to this number using the command `\ref{S:intro}`:

```
In Section~\ref{S:intro}, we introduce ...
```

`\section*` produces an unnumbered section.

Sections have subsections, and subsections have subsubsections, followed by paragraphs and subparagraphs. The corresponding commands are

```
\subsection  \subsubsection  \paragraph  \subparagraph
```

Their unnumbered variants are

```
\subsection*  \subsubsection*  \paragraph*  \subparagraph*
```

### 4.4.3 *Invoking proclamations*

In the preamble of `article.tpl`, you defined the theorem, lemma, proposition, definition, corollary, and notation proclamations. Each of these proclamations defines an environment.

For example, a theorem is typed within a `theorem` environment. The body of the theorem (that is, the part of the source file that produces the theorem) is typed between the lines

```
\begin{theorem}\label{T:xxx}
```

and

```
\end{theorem}
```

where `T:`*xxx* is the label for the theorem. (You should replace *xxx* with a label that is somewhat descriptive of the contents of your theorem.) LaTeX will automatically assign a number to the theorem, and the theorem can be referenced by using a command of the form `\ref{T:`*xxx*`}`.

### 4.4.4 *Inserting references*

Finally, we reach the bibliography. Below are typical entries for the most frequently used types of references: an article in a journal, a book, and a Ph.D. thesis. For more examples, see the bibliographic template file, `bibl.tpl`, in the `samples` directory.

```
\bibitem{eM57}
  Ernest~T. Moynahan,
  \emph{On a problem of M.H. Stone,}
   Acta Math. Acad. Sci. Hungar. \textbf{8} (1957), 455--460.
\bibitem{gM68}
  George~A. Menuhin,
  \emph{Universal Algebra,}
  D.~van Nostrand, Princeton-Toronto-London-Melbourne, 1968.
\bibitem{sF90}
  Soo-Key Foo,
  \emph{Lattice Constructions,} Ph.D. thesis,
  University of Winnebago, Winnebago, MN, December, 1990.
```

This bibliography uses the convention that the label for the `\bibitem` consists of the initials of the author and the year of publication: A publication by Andrew B. Reich in 1987 would have the label aR87 (his second publication from that year would be aR87a). For joint publications, the label consists of the initials of the authors and the year of publication: A publication by John Bradford and Andrew B.

Reich in 1987 would have the label BR87. A reference to Foo's article would be made with `\cite{sF90}`. Of course, you can use any labels you choose.

In this introductory book, we will not discuss Oren Patashnik's BIBTEX program (in the LaTeX distribution), which is very useful for compiling longer bibliographies and adapting bibliographic references to various bibliographic styles.

### 4.4.5 *Adding graphics*

To include graphics (e.g., drawings, scanned images) in an article, start by saving them in EPS (Encapsulated PostScript) format. The standard method for including a graphics file is to use the graphics package by David Carlisle and Sebastian Rahtz, which is part of the LaTeX distribution (see Section C.1). Place the line

```
\usepackage{graphics}
```

in the preamble. A typical figure is then specified as follows:

```
\begin{figure}
   \includegraphics{file.eps}
   \caption{title}\label{Fi:xxx}
\end{figure}
```

The `\caption` command allows you to specify a caption for the figure. Because of the way LaTeX handles this material, commands used in captions and similar structures (e.g., chapter and section names) may be typeset improperly. To safeguard such commands, use `\protect`, as shown in the next example.

If you want to center the figure, you can enclose the `\includegraphics` command in a `\centerline` command. For instance, the figure on page 85 is included with the commands:

```
\begin{figure}[tbh]
   \centerline{\includegraphics{LatStruct.eps}}
   \caption{The structure of \protect\LaTeX.}\label{Fi:LatStruct}
\end{figure}
```

The optional argument `tbh` tells LaTeX where it should try to place the figure. It consists of one to three characters from the following set:

- `h` here, where the `figure` environment is defined
- `t` at the top of a page
- `b` at the bottom of a page
- `p` on a separate page (maybe with other figures)

LaTeX will try to place the figure as specified. To *force* LaTeX to place the figure *here,* use the argument `h!`; the arguments `t!`, `b!`, and `p!` work similarly.

### *4.4.6 Adding tables*

Typesetting a table with the `tabular` environment is almost the same as typesetting a matrix.

Here is a table, centered with a `center` environment:

| Name | 1 | 2 | 3 |
|---|---|---|---|
| Peter | 2.45 | 34.12 | 1.00 |
| John | 0.00 | 12.89 | 3.71 |
| David | 2.00 | 1.85 | 0.71 |

This table was typed as follows:

```
\begin{center}
   \begin{tabular}{||l|r|r|r||}
      \hline
      Name  &  1    & 2     & 3    \\ \hline
      Peter &  2.45 & 34.12 & 1.00\\ \hline
      John  &  0.00 & 12.89 & 3.71\\ \hline
      David &  2.00 & 1.85  & 0.71\\ \hline
   \end{tabular}
\end{center}
```

`\begin{tabular}` takes an argument consisting of a character `l`, `r`, or `c` (for left, right, or center alignment) for each column, and (optionally) `|` symbols, which tell LaTeX to draw vertical lines at the corresponding positions in the table.

Just as we saw in the discussion of matrices, adjacent columns are separated by `&` and adjacent rows are separated by `\\`. Unlike matrices, if you use the `\hline` command to draw horizontal lines between the rows of your table, you *must* specify the final linebreak before your final `\hline` command.

To add tabular material that is set off from the text, use the `table` environment. It is just like the `figure` environment, except that the caption will read "Table" instead of "Figure."

CHAPTER

5

# *An AMS article*

In this chapter, we will discuss the anatomy of an article that uses the AMS article document class, `amsart`, by examining the sample article `sampart.tex`. Type it in as we discuss the parts of an AMS article, or copy it from the `samples` directory (see page 2).

Ⓐ Since this whole chapter deals with AMS topics, the marginal warnings will be omitted for the remainder of the chapter.

## *5.1 The structure of an AMS article*

An AMS article uses the `amsart` document class. Like a LaTeX article, it is made up of two major parts, the preamble and the body (see Section 4.1). The preamble starts with the following lines (the comment lines are ignored by LaTeX):

```
\documentclass{amsart}
\usepackage{amssymb,latexsym}
```

(The amsmath package is automatically loaded by the `amsart` document class.)

Next come the proclamation definitions (see Section 5.2) and the top matter (see Section 5.4), both of which can be more detailed than those in a standard LaTeX article.

## 5.2 *Proclamation definitions*

In `sampart.tex`, the `amsart` sample article, there are several different proclamations that use a variety of styles with varying degrees of emphasis (see the typeset `sampart.tex` on pages 67–69). These proclamations are defined by the following lines:

```
\theoremstyle{plain}
\newtheorem{theorem}{Theorem}
\newtheorem{corollary}{Corollary}
\newtheorem*{main}{Main~Theorem}
\newtheorem{lemma}{Lemma}
\newtheorem{proposition}{Proposition}

\theoremstyle{definition}
\newtheorem{definition}{Definition}

\theoremstyle{remark}
\newtheorem*{notation}{Notation}

\numberwithin{equation}{section}
```

In the proclamation definition

```
\newtheorem{theorem}{Theorem}
```

the first argument (`theorem`) is the name of the environment that will invoke the theorem, and the second argument (`Theorem`) is the text that will be used when LaTeX typesets your theorem.

LaTeX will number the theorems automatically (unless you define an unnumbered theorem with the `\newtheorem*` command) and set them off from the text with white space above and below. It will format the label (e.g., Theorem 1) as specified by the document class `amsart`. The label will be followed by the theorem itself, emphasized (as specified by `amsart`).

When you type a theorem in your document, you may specify an optional argument to the environment that will appear after the numbered label:

```
\begin{theorem}[The Fuchs-Schmidt Theorem]
   The statement of the theorem.
\end{theorem}
```

will be typeset as

**Theorem 2. (The Fuchs-Schmidt Theorem)** *The statement of the theorem.*

### 5.2.1 Consecutive numbering

If you want to number two sets of proclamations consecutively, you can do so by first defining one, and then using its name as an optional argument to the second. For example, to number the lemmas and propositions in your paper consecutively, you would type the following two lines in the preamble:

```
\newtheorem{lemma}{Lemma}
\newtheorem{proposition}[lemma]{Proposition}
```

Lemmas and propositions will now be consecutively numbered as **Lemma 1, Proposition 2, Proposition 3, Lemma 4**, and so on.

Note that the optional argument of a proclamation definition must be the environment name of a proclamation that has *already been defined.*

### 5.2.2 Numbering within a section

The `\newtheorem` command may also have a different optional argument, which causes LaTeX to number the lemmas within sections. For example,

```
\newtheorem{lemma}{Lemma}[section]
```

will cause lemmas in Section 1 to be numbered like **Lemma 1.1** and **Lemma 1.2**; in Section 2, you would have **Lemma 2.1**, **Lemma 2.2**, and so on.

By giving the `[section]` argument to a proclamation that is referenced in another proclamation (so that they are numbered consecutively), you can get a set of proclamations that are numbered consecutively within a section.

### 5.2.3 Proclamations with style

With the amsthm package (which is automatically loaded by all AMS document classes), you can choose one of three styles for your proclamations by preceding them with the `\theoremstyle{`*`style`*`}` command, where *`style`* is one of the following:

- `plain`, the most emphatic
- `definition`
- `remark`, the least emphatic

Examine the `sampart.tex` sample article (on pages 67–69) to see how the chosen styles affect the typeset proclamations.

## 5.3 Equation numbering

Following the proclamation definitions, the line

```
\numberwithin{equation}{section}
```

causes the equations to be numbered within sections. This is very similar to the `[section]` optional argument of `\newtheorem` in Section 5.2.

## 5.4 Top matter

Examine the top matter of `sampart.tex` on page 70, and then look at page 71 to see how the title page is put together from these pieces.

The title page information is provided as the arguments of several commands. For your convenience, we divide them into three groups.

There is only one general rule:

---

**Rule** ■ AMS top matter
There can be no blank lines in the argument of any AMS top-matter command.

---

### 5.4.1 Article information

**Commands:**

- `\title`: The article's title. Specify linebreaks with the `\\` command. Do not put a period at the end of a title. An optional argument is a short title for use in running heads.
- `\translator`: The translator of the article (optional). Specify linebreaks with the `\\` command. Do not add a period to the end of the argument.
- `\dedicatory`: For use in dedicating an article (optional). Specify linebreaks with the `\\` command.
- `\date`: The date. Use
  - `\date{\today}` to get the date on which the document is typeset
  - `\date{`*`date`*`}` for a specific date, for example, `\date{March 15, 1999}`
  - `\date{}` for no date (or omit the `\date` command)

  Note that this command behaves differently in an AMS article than in a LaTeX article.

### 5.4.2 Author information

**Commands:**

- `\author`: The author's name. An optional argument (between `\author` and the author's name in braces) is a short form of the author's name for use in running heads (e.g., `\author[F.L. Stevens]{Franklin~L. Stevens}`).
- `\address`: The author's address. Separate lines with the `\\` command. The optional argument is the name of the author to whom this address applies; use this in articles with multiple authors.

- `\curraddress`: Current address. Separate lines with the `\\` command. The optional argument is the name of the author to whom this address applies. (As for the `\address`.)
- `\thanks`: Research support or other acknowledgment. Do not specify linebreaks.
- `\email`: E-mail address. The optional argument is the name of the author to whom this address applies.
- `\urladdr`: URL for the author's Web page.

### 5.4.3 *AMS information*

**Commands:**

- `\subjclass`: The AMS subject classification. Enter as
  `\subjclass{Primary` *primary*; `Secondary` *secondary*`}`
  You must specify the full numbers: See the AMS subject classification at
  `http://www.ams.org/index/msc/MSC.html`
- `\keywords`: The `amsart` document class supplies the phrase
  *Key words and phrases.*
  and a period after your keywords. This command is optional.

The arguments of these two commands (along with the arguments of `\thanks` and `\date`) are collected and typeset together at the bottom of the first page of the article as unmarked footnotes.

For an example of the author information for an article with a single author, see the first page of the source file of `sampart.tex` on page 70. Here is an example of the author information section from an article with two authors:

```
% Author information
\author{George~A. Menuhin}
\address{Computer Science Department\\
        University of Winnebago\\
        Winnebago, Minnesota 53714}
\email{gmen@ccw.uwinnebago.edu}
\urladdr{http://math.uwinnebago.ca/homepages/menuhin/}
\thanks{The research of the first author was
       supported by the NSF under grant number~23466.}
\author{Ernest~T. Moynahan}
\address{Mathematical Research Institute
        of the Hungarian Academy of Sciences\\
        Budapest, P.O.B. 127, H-1364\\
        Hungary}
\email{moynahan@math.bme.hu}
\urladdr{http://www.math.bme.hu/~moynahan/}
\thanks{The research of the second author
```

```
was supported by the Hungarian
National Foundation for Scientific Research,
under Grant No.~9901.}
```

You can create AMS article templates using the amsart document class for your own use, just as you did for LaTeX articles in Section 4.3. An example is provided in the samples directory: ggamsart.tpl is a template personalized for my use.

## 5.5 Proofs

The amsthm package also defines a proof environment. For example,

*Proof.* This is a proof, delimited by the q.e.d. symbol. □

typed as

```
\begin{proof}
  This is a proof, delimited by the q.e.d. symbol.
\end{proof}
```

## 5.6 The AMS sample article

sampart.tex is the source file for our sample article using the AMS article document class, amsart.

The typeset sampart.tex is printed on the following three pages.

On the subsequent seven pages, the source file and the typeset version of the AMS sample article, sampart.tex, are shown together, so you can see how the marked-up source file becomes the typeset article.

# A CONSTRUCTION OF COMPLETE-SIMPLE DISTRIBUTIVE LATTICES

GEORGE A. MENUHIN

ABSTRACT. In this note we prove that there exist *complete-simple distributive lattices,* that is, complete distributive lattices in which there are only two complete congruences.

## 1. INTRODUCTION

In this note we prove the following result:

**Main Theorem.** *There exists an infinite complete distributive lattice $K$ with only the two trivial complete congruence relations.*

## 2. THE $D^{\langle 2 \rangle}$ CONSTRUCTION

For the basic notation in lattice theory and universal algebra, see Ferenc R. Richardson [5] and George A. Menuhin [2]. We start with some definitions:

**Definition 1.** Let $V$ be a complete lattice, and let $\mathfrak{p} = [u, v]$ be an interval of $V$. Then $\mathfrak{p}$ is called *complete-prime* if the following three conditions are satisfied:

(1) $u$ is meet-irreducible but $u$ is *not* completely meet-irreducible;
(2) $v$ is join-irreducible but $v$ is *not* completely join-irreducible;
(3) $[u, v]$ is a complete-simple lattice.

Now we prove the following result:

**Lemma 1.** *Let $D$ be a complete distributive lattice satisfying conditions* (1) *and* (2). *Then $D^{\langle 2 \rangle}$ is a sublattice of $D^2$; hence $D^{\langle 2 \rangle}$ is a lattice, and $D^{\langle 2 \rangle}$ is a complete distributive lattice satisfying conditions* (1) *and* (2).

*Proof.* By conditions (1) and (2), $D^{\langle 2 \rangle}$ is a sublattice of $D^2$. Hence, $D^{\langle 2 \rangle}$ is a lattice.

Since $D^{\langle 2 \rangle}$ is a sublattice of a distributive lattice, $D^{\langle 2 \rangle}$ is a distributive lattice. Using the characterization of standard ideals in Ernest T. Moynahan [3], $D^{\langle 2 \rangle}$ has a zero and a unit element, namely, $\langle 0, 0 \rangle$ and $\langle 1, 1 \rangle$. To show that $D^{\langle 2 \rangle}$ is complete, let $\varnothing \neq A \subseteq D^{\langle 2 \rangle}$, and let $a = \bigvee A$ in $D^2$. If $a \in D^{\langle 2 \rangle}$, then $a = \bigvee A$ in $D^{\langle 2 \rangle}$; otherwise, $a$ is of the form $\langle b, 1 \rangle$ for some $b \in D$ with $b < 1$. Now $\bigvee A = \langle 1, 1 \rangle$ in $D^2$ and the dual argument shows that $\bigwedge A$ also exists in $D^2$. Hence $D$ is complete. Conditions (1) and (2) are obvious for $D^{\langle 2 \rangle}$. □

**Corollary 1.** *If $D$ is complete-prime, then so is $D^{\langle 2 \rangle}$.*

---

*Date*: March 15, 1999.

1991 *Mathematics Subject Classification.* Primary: 06B10; Secondary: 06D05.

*Key words and phrases.* Complete lattice, distributive lattice, complete congruence, congruence lattice.

Research supported by the NSF under grant number 23466.

The motivation for the following result comes from Soo-Key Foo [1].

**Lemma 2.** *Let $\Theta$ be a complete congruence relation of $D^{\langle 2 \rangle}$ such that*

$$\langle 1, d \rangle \equiv \langle 1, 1 \rangle \pmod{\Theta}, \tag{2.1}$$

*for some $d \in D$ with $d < 1$. Then $\Theta = \iota$.*

*Proof.* Let $\Theta$ be a complete congruence relation of $D^{\langle 2 \rangle}$ satisfying (2.1). Then $\Theta = \iota$. □

## 3. The $\Pi^*$ construction

The following construction is crucial to our proof of the Main Theorem:

**Definition 2.** Let $D_i$, for $i \in I$, be complete distributive lattices satisfying condition (2). Their $\Pi^*$ product is defined as follows:

$$\Pi^*(D_i \mid i \in I) = \Pi(D_i^- \mid i \in I) + 1;$$

that is, $\Pi^*(D_i \mid i \in I)$ is $\Pi(D_i^- \mid i \in I)$ with a new unit element.

*Notation.* If $i \in I$ and $d \in D_i^-$, then

$$\langle \dots, 0, \dots, \overset{i}{d}, \dots, 0, \dots \rangle$$

is the element of $\Pi^*(D_i \mid i \in I)$ whose $i$-th component is $d$ and all the other components are 0.

See also Ernest T. Moynahan [4]. Next we verify:

**Theorem 1.** *Let $D_i$, for $i \in I$, be complete distributive lattices satisfying condition (2). Let $\Theta$ be a complete congruence relation on $\Pi^*(D_i \mid i \in I)$. If there exist $i \in I$ and $d \in D_i$ with $d < 1_i$ such that for all $d \leq c < 1_i$,*

$$\langle \dots, 0, \dots, \overset{i}{d}, \dots, 0, \dots \rangle \equiv \langle \dots, 0, \dots, \overset{i}{c}, \dots, 0, \dots \rangle \pmod{\Theta}, \tag{3.1}$$

*then $\Theta = \iota$.*

*Proof.* Since

$$\langle \dots, 0, \dots, \overset{i}{d}, \dots, 0, \dots \rangle \equiv \langle \dots, 0, \dots, \overset{i}{c}, \dots, 0, \dots \rangle \pmod{\Theta}, \tag{3.2}$$

and $\Theta$ is a complete congruence relation, it follows from condition (3) that

$$\begin{aligned} \langle \dots, \overset{i}{d}, \dots, 0, \dots \rangle &\equiv \\ &\bigvee (\langle \dots, 0, \dots, \overset{i}{c}, \dots, 0, \dots \rangle \mid d \leq c < 1) \equiv 1 \pmod{\Theta}. \end{aligned} \tag{3.3}$$

Let $j \in I$ for $j \neq i$, and let $a \in D_j^-$. Meeting both sides of the congruence (3.2) with $\langle \dots, 0, \dots, \overset{j}{a}, \dots, 0, \dots \rangle$, we obtain

$$\begin{aligned} 0 = \langle \dots, 0, \dots, \overset{i}{d}, \dots, 0, \dots \rangle \wedge \langle \dots, 0, \dots, \overset{j}{a}, \dots, 0, \dots \rangle \\ \equiv \langle \dots, 0, \dots, \overset{j}{a}, \dots, 0, \dots \rangle \pmod{\Theta}. \end{aligned} \tag{3.4}$$

Using the completeness of $\Theta$ and (3.4), we get:

$$0 \equiv \bigvee (\langle \dots, 0, \dots, \overset{j}{a}, \dots, 0, \dots \rangle \mid a \in D_j^-) = 1 \pmod{\Theta},$$

hence $\Theta = \iota$. □

**Theorem 2.** *Let $D_i$ for $i \in I$ be complete distributive lattices satisfying conditions* (2) *and* (3). *Then $\Pi^*(D_i \mid i \in I)$ also satisfies conditions* (2) *and* (3).

*Proof.* Let $\Theta$ be a complete congruence on $\Pi^*(D_i \mid i \in I)$. Let $i \in I$. Define

$$\widehat{D}_i = \{\langle \ldots, 0, \ldots, \overset{i}{d}, \ldots, 0, \ldots \rangle \mid d \in D_i^- \} \cup \{1\}.$$

Then $\widehat{D}_i$ is a complete sublattice of $\Pi^*(D_i \mid i \in I)$, and $\widehat{D}_i$ is isomorphic to $D_i$. Let $\Theta_i$ be the restriction of $\Theta$ to $\widehat{D}_i$.

Since $D_i$ is complete-simple, so is $\widehat{D}_i$, and hence $\Theta_i$ is $\omega$ or $\iota$. If $\Theta_i = \rho$ for all $i \in I$, then $\Theta = \omega$. If there is an $i \in I$, such that $\Theta_i = \iota$, then $0 \equiv 1 \pmod{\Theta}$, hence $\Theta = \iota$. □

The Main Theorem follows easily from Theorems 1 and 2.

## References


[1] Soo-Key Foo, *Lattice Constructions*, Ph.D. thesis, University of Winnebago, Winnebago, MN, December, 1990.

[2] George A. Menuhin, *Universal Algebra*, D. van Nostrand, Princeton-Toronto-London-Melbourne, 1968.

[3] Ernest T. Moynahan, *On a problem of M.H. Stone*, Acta Math. Acad.Sci. Hungar. **8** (1957), 455–460.

[4] ———, *Ideals and congruence relations in lattices. II*, Magyar Tud. Akad. Mat. Fiz. Oszt. Közl. **9** (1957), 417–434 (Hungarian).

[5] Ferenc R. Richardson, *General Lattice Theory*, Mir, Moscow, expanded and revised ed., 1982 (Russian).



Computer Science Department, University of Winnebago, Winnebago, Minnesota 53714

*E-mail address*: `menuhin@ccw.uwinnebago.edu`

*URL*: `http://math.uwinnebago.ca/homepages/menuhin/`

```
% Sample file: sampart.tex
% The sample article for the amsart document class
% Typeset with LaTeX format

\documentclass{amsart}
\usepackage{amssymb,latexsym}
\theoremstyle{plain}
\newtheorem{theorem}{Theorem}
\newtheorem{corollary}{Corollary}
\newtheorem*{main}{Main~Theorem}
\newtheorem{lemma}{Lemma}
\newtheorem{proposition}{Proposition}
\theoremstyle{definition}
\newtheorem{definition}{Definition}
\theoremstyle{remark}
\newtheorem*{notation}{Notation}
\numberwithin{equation}{section}

\begin{document}
\title[Complete-simple distributive lattices]
      {A construction of complete-simple\\
       distributive lattices}
\author{George~A. Menuhin}
\address{Computer Science Department\\
         University of Winnebago\\
         Winnebago, Minnesota 53714}
\email{menuhin@ccw.uwinnebago.edu}
\urladdr{http://math.uwinnebago.ca/homepages/menuhin/}
\thanks{Research supported by the NSF under grant number~23466.}
\keywords{Complete lattice, distributive lattice,
          complete congruence, congruence lattice}
\subjclass{Primary: 06B10; Secondary: 06D05}
\date{March 15, 1999}
\begin{abstract}
   In this note we prove that there exist \emph{complete-simple
   distributive lattices,} that is, complete distributive
   lattices in which there are only two complete congruences.
\end{abstract}

\maketitle

\section{Introduction}\label{S:intro}
In this note we prove the following result:

\begin{main}
   There exists an infinite complete distributive lattice $K$
   with only the two trivial complete congruence relations.
\end{main}

\section{The $D^{\langle 2 \rangle}$ construction}\label{S:Ds}
For the basic notation in lattice theory and universal algebra,
see Ferenc~R. Richardson~\cite{fR82} and
George~A. Menuhin~\cite{gM68}. We start with some definitions:
```

# A CONSTRUCTION OF COMPLETE-SIMPLE DISTRIBUTIVE LATTICES

GEORGE A. MENUHIN

ABSTRACT. In this note we prove that there exist *complete-simple distributive lattices,* that is, complete distributive lattices in which there are only two complete congruences.

## 1. INTRODUCTION

In this note we prove the following result:

**Main Theorem.** *There exists an infinite complete distributive lattice $K$ with only the two trivial complete congruence relations.*

## 2. THE $D^{\langle 2 \rangle}$ CONSTRUCTION

For the basic notation in lattice theory and universal algebra, see Ferenc R. Richardson [5] and George A. Menuhin [2]. We start with some definitions:

*Date*: March 15, 1999.

1991 *Mathematics Subject Classification.* Primary: 06B10; Secondary: 06D05.

*Key words and phrases.* Complete lattice, distributive lattice, complete congruence, congruence lattice.

Research supported by the NSF under grant number 23466.

**Definition 1.** Let $V$ be a complete lattice, and let $\mathfrak{p} = [u, v]$ be an interval of $V$. Then $\mathfrak{p}$ is called *complete-prime* if the following three conditions are satisfied:

(1) $u$ is meet-irreducible but $u$ is *not* completely meet-irreducible;
(2) $v$ is join-irreducible but $v$ is *not* completely join-irreducible;
(3) $[u, v]$ is a complete-simple lattice.

Now we prove the following result:

**Lemma 1.** *Let $D$ be a complete distributive lattice satisfying conditions* (1) *and* (2). *Then $D^{\langle 2 \rangle}$ is a sublattice of $D^{2}$; hence $D^{\langle 2 \rangle}$ is a lattice, and $D^{\langle 2 \rangle}$ is a complete distributive lattice satisfying conditions* (1) *and* (2).

*Proof.* By conditions (1) and (2), $D^{\langle 2 \rangle}$ is a sublattice of $D^{2}$. Hence, $D^{\langle 2 \rangle}$ is a lattice.

```
\begin{definition}\label{D:prime}
   Let $V$ be a complete lattice, and let $\mathfrak{p} = [u, v]$ be
   an interval of $V$.  Then $\mathfrak{p}$ is called
   \emph{complete-prime} if the following three conditions are satisfied:
   \begin{itemize}
      \item[(1)] $u$ is meet-irreducible but $u$ is \emph{not}
         completely meet-irreducible;
      \item[(2)] $v$ is join-irreducible but $v$ is \emph{not}
         completely join-irreducible;
      \item[(3)] $[u, v]$ is a complete-simple lattice.
   \end{itemize}
\end{definition}

Now we prove the following result:

\begin{lemma}\label{L:ds}
   Let $D$ be a complete distributive lattice satisfying
   conditions~\textup{(1)} and~\textup{(2)}.  Then
   $D^{\langle 2 \rangle}$ is a sublattice of $D^{2}$;
   hence $D^{\langle 2 \rangle}$ is a lattice, and
   $D^{\langle 2 \rangle}$ is a complete distributive
   lattice satisfying conditions~\textup{(1)} and \textup{(2)}.
\end{lemma}

\begin{proof}
   By conditions~(1) and (2), $D^{\langle 2 \rangle}$ is a sublattice
   of $D^{2}$.  Hence, $D^{\langle 2 \rangle}$ is a lattice.
```

Since $D^{\langle 2 \rangle}$ is a sublattice of a distributive lattice, $D^{\langle 2 \rangle}$ is a distributive lattice. Using the characterization of standard ideals in Ernest T. Moynahan [3], $D^{\langle 2 \rangle}$ has a zero and a unit element, namely, $\langle 0, 0 \rangle$ and $\langle 1, 1 \rangle$. To show that $D^{\langle 2 \rangle}$ is complete, let $\varnothing \ne A \subseteq D^{\langle 2 \rangle}$, and let $a = \bigvee A$ in $D^{2}$. If $a \in D^{\langle 2 \rangle}$, then $a = \bigvee A$ in $D^{\langle 2 \rangle}$; otherwise, $a$ is of the form $\langle b, 1 \rangle$ for some $b \in D$ with $b < 1$. Now $\bigvee A = \langle 1, 1\rangle$ in $D^{2}$ and the dual argument shows that $\bigwedge A$ also exists in $D^{2}$. Hence $D$ is complete. Conditions (1) and (2) are obvious for $D^{\langle 2 \rangle}$. □

**Corollary 1.** *If $D$ is complete-prime, then so is $D^{\langle 2 \rangle}$.*

The motivation for the following result comes from Soo-Key Foo [1].

**Lemma 2.** *Let $\Theta$ be a complete congruence relation of $D^{\langle 2 \rangle}$ such that*

$$\langle 1, d \rangle \equiv \langle 1, 1 \rangle \pmod{\Theta}, \tag{2.1}$$

*for some $d \in D$ with $d < 1$. Then $\Theta = \iota$.*

```
   Since $D^{\langle 2 \rangle}$ is a sublattice of a distributive
   lattice, $D^{\langle 2 \rangle}$ is a distributive lattice.  Using
   the characterization of standard ideals in Ernest~T.
   Moynahan~\cite{eM57}, $D^{\langle 2 \rangle}$ has a zero and a unit
   element, namely, $\langle 0, 0 \rangle$ and $\langle 1, 1 \rangle$.
   To show that $D^{\langle 2 \rangle}$ is complete, let
   $\varnothing \ne A \subseteq D^{\langle 2 \rangle}$, and let
   $a = \bigvee A$ in $D^{2}$.  If
   $a \in D^{\langle 2 \rangle}$, then
   $a = \bigvee A$ in $D^{\langle 2 \rangle}$; otherwise, $a$
   is of the form $\langle b, 1 \rangle$ for some
   $b \in D$ with $b < 1$.  Now $\bigvee A = \langle 1, 1\rangle$
   in $D^{2}$ and the dual argument shows that $\bigwedge A$ also
   exists in $D^{2}$.  Hence $D$ is complete. Conditions~(1) and
   (2) are obvious for $D^{\langle 2 \rangle}$.
\end{proof}

\begin{corollary}\label{C:prime}
   If $D$ is complete-prime, then so is $D^{\langle 2 \rangle}$.
\end{corollary}

The motivation for the following result comes from Soo-Key Foo~\cite{sF90}.

\begin{lemma}\label{L:ccr}
   Let $\Theta$ be a complete congruence relation of
   $D^{\langle 2 \rangle}$ such that
   \begin{equation}\label{E:rigid}
      \langle 1, d \rangle \equiv \langle 1, 1 \rangle \pmod{\Theta},
   \end{equation}
   for some $d \in D$ with $d < 1$. Then $\Theta = \iota$.
\end{lemma}
```

*Proof.* Let $\Theta$ be a complete congruence relation of $D^{\langle 2 \rangle}$ satisfying (2.1). Then $\Theta = \iota$. □

## 3. The $\Pi^*$ construction

The following construction is crucial to our proof of the Main Theorem:

**Definition 2.** Let $D_i$, for $i \in I$, be complete distributive lattices satisfying condition (2). Their $\Pi^*$ product is defined as follows:

$$\Pi^*(D_i \mid i \in I) = \Pi(D_i^- \mid i \in I) + 1;$$

that is, $\Pi^*(D_i \mid i \in I)$ is $\Pi(D_i^- \mid i \in I)$ with a new unit element.

*Notation.* If $i \in I$ and $d \in D_i^-$, then

$$\langle \ldots, 0, \ldots, \overset{i}{d}, \ldots, 0, \ldots \rangle$$

is the element of $\Pi^*(D_i \mid i \in I)$ whose $i$-th component is $d$ and all the other components are 0.

```
\begin{proof}
   Let $\Theta$ be a complete congruence relation of
   $D^{\langle 2 \rangle}$ satisfying \eqref{E:rigid}. Then $\Theta =
\iota$.
\end{proof}

\section{The $\Pi^{*}$ construction}\label{S:P*}
The following construction is crucial to our proof of the Main Theorem:

\begin{definition}\label{D:P*}
   Let $D_{i}$, for $i \in I$, be complete distributive lattices
   satisfying condition~\textup{(2)}.  Their $\Pi^{*}$ product is defined
as
   follows:
   \[
      \Pi^{*} ( D_{i} \mid i \in I ) = \Pi ( D_{i}^{-} \mid i \in I ) + 1;
   \]
   that is, $\Pi^{*} ( D_{i} \mid i \in I )$ is $\Pi ( D_{i}^{-} \mid
   i \in I )$ with a new unit element.
\end{definition}

\begin{notation}
   If $i \in I$ and $d \in D_{i}^{-}$, then
   \[
      \langle \ldots, 0, \ldots, \overset{i}{d}, \ldots, 0, \ldots \rangle
   \]
   is the element of $\Pi^{*} ( D_{i} \mid i \in I )$ whose $i$-th
   component is $d$ and all the other components are $0$.
\end{notation}
```

See also Ernest T. Moynahan [4]. Next we verify:

**Theorem 1.** *Let $D_i$, for $i \in I$, be complete distributive lattices satisfying condition* (2). *Let $\Theta$ be a complete congruence relation on $\Pi^*(D_i \mid i \in I)$. If there exist $i \in I$ and $d \in D_i$ with $d < 1_i$ such that for all $d \leq c < 1_i$,*

$$\langle \ldots, 0, \ldots, \overset{i}{d}, \ldots, 0, \ldots \rangle \equiv \langle \ldots, 0, \ldots, \overset{i}{c}, \ldots, 0, \ldots \rangle \pmod{\Theta}, \tag{3.1}$$

*then $\Theta = \iota$.*

*Proof.* Since

$$\langle \ldots, 0, \ldots, \overset{i}{d}, \ldots, 0, \ldots \rangle \equiv \langle \ldots, 0, \ldots, \overset{i}{c}, \ldots, 0, \ldots \rangle \pmod{\Theta}, \tag{3.2}$$

and $\Theta$ is a complete congruence relation, it follows from condition (3) that

$$\begin{aligned} & \langle \ldots, \overset{i}{d}, \ldots, 0, \ldots \rangle \equiv \\ &\qquad \quad \bigvee ( \langle \ldots, 0, \ldots, \overset{i}{c}, \ldots, 0, \ldots \rangle \mid d \leq c < 1 ) \equiv 1 \pmod{\Theta}. \end{aligned} \tag{3.3}$$

Let $j \in I$ for $j \neq i$, and let $a \in D_{j}^{-}$. Meeting both sides of the congruence (3.2) with $\langle \ldots, 0, \ldots, \overset{j}{a}, \ldots, 0, \ldots \rangle$, we obtain

```
See also Ernest~T. Moynahan \cite{eM57a}.  Next we verify:
\begin{theorem}\label{T:P*}
   Let $D_{i}$, for $i \in I$, be complete distributive lattices satisfying
   condition~\textup{(2)}.  Let $\Theta$ be a complete congruence relation on
   $\Pi^{*} ( D_{i} \mid i \in I)$.  If there exist $i \in I$ and $d \in D_{i}$
   with $d < 1_{i}$ such that for all $d \leq c < 1_{i}$,
   \begin{equation}\label{E:cong1}
      \langle \ldots, 0, \ldots,\overset{i}{d},
      \ldots, 0, \ldots \rangle \equiv \langle \ldots, 0, \ldots,
      \overset{i}{c}, \ldots, 0, \ldots \rangle \pmod{\Theta},
   \end{equation}
   then $\Theta = \iota$.
\end{theorem}
\begin{proof}
   Since
   \begin{equation}\label{E:cong2}
      \langle \ldots, 0, \ldots, \overset{i}{d}, \ldots, 0,
         \ldots \rangle \equiv \langle \ldots, 0, \ldots,
         \overset{i}{c}, \ldots, 0, \ldots \rangle \pmod{\Theta},
   \end{equation}
   and $\Theta$ is a complete congruence relation, it follows from condition~(3) that
   \begin{align}\label{E:cong}
      & \langle \ldots, \overset{i}{d}, \ldots, 0, \ldots \rangle \equiv\\
      &\qquad \quad \bigvee ( \langle \ldots, 0, \ldots, \overset{i}{c},
       \ldots, 0, \ldots \rangle \mid d \leq c < 1 ) \equiv 1 \pmod{\Theta}. \notag
   \end{align}

   Let $j \in I$ for $j \neq i$, and let $a \in D_{j}^{-}$. Meeting both sides of the
   congruence \eqref{E:cong2} with $\langle \ldots, 0, \ldots, \overset{j}{a}, \ldots,
   0,\ldots \rangle$, we obtain
```

$$0 = \langle \ldots, 0, \ldots, \overset{i}{d}, \ldots, 0, \ldots \rangle \wedge \langle \ldots, 0, \ldots, \overset{j}{a}, \ldots, 0, \ldots \rangle \tag{3.4}$$
$$\equiv \langle \ldots, 0, \ldots, \overset{j}{a}, \ldots, 0, \ldots \rangle \pmod{\Theta}.$$

Using the completeness of $\Theta$ and (3.4), we get:

$$0 \equiv \bigvee ( \langle \ldots, 0, \ldots, \overset{j}{a}, \ldots, 0, \ldots \rangle \mid a \in D_{j}^{-} ) = 1 \pmod{\Theta},$$

hence $\Theta = \iota$. □

**Theorem 2.** *Let $D_i$ for $i \in I$ be complete distributive lattices satisfying conditions* (2) *and* (3). *Then $\Pi^*(D_i \mid i \in I)$ also satisfies conditions* (2) *and* (3).

*Proof.* Let $\Theta$ be a complete congruence on $\Pi^*(D_i \mid i \in I)$. Let $i \in I$. Define

$$\widehat{D}_{i} = \{ \langle \ldots, 0, \ldots, \overset{i}{d}, \ldots, 0, \ldots \rangle \mid d \in D_{i}^{-} \} \cup \{ 1 \}.$$

Then $\widehat{D}_i$ is a complete sublattice of $\Pi^*(D_i \mid i \in I)$, and $\widehat{D}_i$ is isomorphic to $D_i$. Let $\Theta_i$ be the restriction of $\Theta$ to $\widehat{D}_i$.

```
   \begin{align}\label{E:comp}
      0 &= \langle \ldots, 0, \ldots, \overset{i}{d}, \ldots, 0, \ldots \rangle \wedge
           \langle \ldots, 0, \ldots, \overset{j}{a}, \ldots, 0, \ldots \rangle\\
        &\equiv \langle \ldots, 0, \ldots, \overset{j}{a}, \ldots, 0, \ldots \rangle
           \pmod{\Theta}.\notag
   \end{align}
   Using the completeness of $\Theta$ and \eqref{E:comp}, we get:
   \[
      0 \equiv \bigvee ( \langle \ldots, 0, \ldots, \overset{j}{a},
      \ldots, 0, \ldots \rangle \mid a \in D_{j}^{-} ) = 1 \pmod{\Theta},
   \]
   hence $\Theta = \iota$.
\end{proof}

\begin{theorem}\label{T:P*a}
   Let $D_{i}$ for $i \in I$ be complete distributive lattices satisfying conditions
   \textup{(2)} and \textup{(3)}.  Then $\Pi^{*} ( D_{i} \mid i \in I )$ also
   satisfies conditions \textup{(2)} and \textup{(3)}.
\end{theorem}

\begin{proof}
   Let $\Theta$ be a complete congruence on
   $\Pi^{*} ( D_{i} \mid i \in I )$. Let $i \in I$.  Define
   \[
      \widehat{D}_{i} = \{ \langle \ldots, 0, \ldots, \overset{i}{d},
      \ldots, 0, \ldots \rangle \mid d \in D_{i}^{-} \} \cup \{ 1 \}.
   \]
   Then $\widehat{D}_{i}$ is a complete sublattice of  $\Pi^{*}
   ( D_{i} \mid i \in I)$, and $\widehat{D}_{i}$ is isomorphic to $D_{i}$.
   Let $\Theta_{i}$ be the restriction of $\Theta$ to $\widehat{D}_{i}$.
```

Since $D_i$ is complete-simple, so is $\widehat{D}_i$, and hence $\Theta_i$ is $\omega$ or $\iota$. If $\Theta_i = \rho$ for all $i \in I$, then $\Theta = \omega$. If there is an $i \in I$, such that $\Theta_i = \iota$, then $0 \equiv 1 \pmod{\Theta}$, hence $\Theta = \iota$. □

The Main Theorem follows easily from Theorems 1 and 2.

## References

[1] Soo-Key Foo, *Lattice Constructions,* Ph.D. thesis, University of Winnebago, Winnebago, MN, December, 1990.

[2] George A. Menuhin, *Universal Algebra,* D. van Nostrand, Princeton-Toronto-London-Melbourne, 1968.

[3] Ernest T. Moynahan, *On a problem of M.H. Stone,* Acta Math. Acad.Sci. Hungar. **8** (1957), 455–460.

[4] ———, *Ideals and congruence relations in lattices. II,* Magyar Tud. Akad. Mat. Fiz. Oszt. Közl. **9** (1957), 417–434 (Hungarian).

[5] Ferenc R. Richardson, *General Lattice Theory,* Mir, Moscow, expanded and revised ed., 1982 (Russian).

Computer Science Department, University of Winnebago, Winnebago, Minnesota 53714

*E-mail address*: menuhin@ccw.uwinnebago.edu

*URL*: http://math.uwinnebago.ca/homepages/menuhin/

```
   Since $D_{i}\) is complete-simple, so is $\widehat{D}_{i}$, and hence $\Theta_{i}$
   is $\omega$ or $\iota$. If $\Theta_{i} = \rho$ for all $i \in I$, then
   $\Theta = \omega$.  If there is an $i \in I$, such that $\Theta_{i} = \iota$,
   then $0 \equiv 1 \pmod{\Theta}$, hence $\Theta = \iota$.
\end{proof}

The Main Theorem follows easily from Theorems~\ref{T:P*} and \ref{T:P*a}.

\begin{thebibliography}{9}
   \bibitem{sF90}
      Soo-Key Foo, \emph{Lattice Constructions,} Ph.D. thesis, University
      of Winnebago, Winnebago, MN, December, 1990.
   \bibitem{gM68}
      George~A. Menuhin, \emph{Universal Algebra,} D.~van Nostrand,
      Princeton-Toronto-London-Mel\-bourne, 1968.
   \bibitem{eM57}
      Ernest~T. Moynahan, \emph{On a problem of M.H. Stone,} Acta Math.
       Acad.Sci. Hungar. \textbf{8} (1957), 455--460.
   \bibitem{eM57a}
      \bysame, \emph{Ideals and congruence relations in lattices.~II,}
     Magyar Tud. Akad. Mat. Fiz. Oszt. K\"{o}zl. \textbf{9} (1957),
     417--434  (Hungarian).
   \bibitem{fR82}
      Ferenc~R. Richardson, \emph{General Lattice Theory,} Mir, Moscow,
      expanded and revised ed., 1982 (Russian).
\end{thebibliography}
\end{document}
```

Note that this sample article is much simpler than most real articles, which almost always contain a number of user-defined commands. In fact, it is customary to collect your most frequently used user-defined commands into a *style file,* a file with a `.sty` extension, and include this file in your source files with a `\usepackage` command, usually after all of the other `\usepackage` commands. User-defined commands that are specific to one article are usually placed in the preamble of that article.

**`sampart2.tex`** is **`sampart.tex`** rewritten in this form; the style file is named **`lattice.sty`**. You will find both files in the `samples` directory.

CHAPTER

6

# Working with LATEX

We conclude this introductory book by discussing the process of working with LATEX, including LATEX error messages, the distinction between logical and visual design, spell checkers and text editors, LATEX for non-English languages, and some reading material.

## 6.1 LATEX error messages

You will probably make a number of mistakes in your first article. These mistakes will fall into one of the following categories:

1. Typographical errors, which LATEX will blindly typeset.
2. Errors in mathematical formulas or in the formatting of the text.
3. Errors in your instructions to LATEX (commands and environments).

Typographical errors can be corrected by viewing the typeset article, finding the errors, and then editing the source file. Using a spell checker before typesetting will help catch many of these errors: See Section 6.4.3 for more information.

Mistakes in the second and third categories will probably trigger errors during the typesetting process (we looked at a few math errors in Section 2.2), some of which will require correction before your article can be completely typeset.

We will now look at some examples of this class of errors by deliberately introducing a number of mistakes into the source file of the introductory LaTeX sample article, `intrart.tex` (in your `work` directory, source file on pages 48–52, and shown typeset on pages 53–54), and examining the error messages that occur.

When LaTeX displays a `?` prompt, you can either try to continue typesetting the document by pressing Return, or type `x` to stop typesetting immediately. See Section 6.4.1 for other options.

**Example 1** In `intrart.tex`, go to line 21 (avoid counting lines by using your editor's "go to line" function or searching for some text) and remove the closing brace so that it reads

```
\begin{abstract
```

When you typeset `intrart.tex`, LaTeX reports a problem:

```
Runaway argument?
{abstract In this note, we prove that there exist \emph \ETC.
! Paragraph ended before \begin was complete.
<to be read again>
                   \par
l.26
```

Line 26 of the file is the line after `\end{abstract}`. From the error message, you can tell that something is wrong with the `abstract` environment.

**Example 2** Now correct line 21, then go to line 25 and change it from

```
\end{abstract}
```

to

```
\end{abstrac}
```

and typeset the article again. LaTeX will inform you of another error:

```
! LaTeX Error: \begin{abstract} on input line 21
  ended by \end{abstrac}.

See the LaTeX manual or LaTeX Companion for explanation.
Type  H <return>  for immediate help.
 ...

l.25 \end{abstrac}
```

You may continue typesetting the article by pressing Return: LaTeX will recover from this error.

**Example 3** Instead of correcting the error in line 25, comment it out:

```
% \end{abstrac}
```

Introduce an additional error in line 66. This line initially reads:

```
   lattices satisfying condition~\textup{(J)}.  Let $\Theta$
```

Change `\Theta` to `\Teta`:

```
   lattices satisfying condition~\textup{(J)}.  Let $\Teta$
```

Now, when you typeset the article, LaTeX reports

```
! Undefined control sequence.
<recently read> \Teta

l.66 ...textup{(J)}.  Let $\Teta
                                $
```

and pressing Return results in the message

```
! LaTeX Error: \begin{abstract} on input line 21
               ended by \end{document}.

See the LaTeX manual or LaTeX Companion for explanation.
Type  H <return>  for immediate help.
 ...

l.126 \end{document}
```

These two mistakes are easy to identify: `\Teta` is a misspelling of `\Theta`, and LaTeX tries to match

```
\begin{abstract}
```

with

```
\end{document}
```

Now undo the two changes you made (uncommenting line 25 and replacing the "`t`" at the end of the word "`abstract`"; replacing the "`h`" to correct the error on line 66).

**Example 4** In line 38, drop the closing brace of the `\label` command:

```
\begin{definition}\label{D:P*
```

This results in the message

```
Runaway argument?
{D:P* Let $D_{i}$, for $i \in I$, be complete distribu\ETC.
! Paragraph ended before \label was complete.
<to be read again>
                   \par
l.49
```

Line 49 is the blank line following `\end{definition}`. The error message is easy to understand: You cannot begin a new paragraph (`\par`) within the argument of a `\label` command.

Undo the change you made to line 38.

**Example 5** Add a blank line following line 53:

```
\langle \ldots, 0, \ldots, d, \ldots, 0, \ldots \rangle
```

This results in the message

```
! Missing $ inserted.
<inserted text>
                $
l.54
```

There can be no blank lines within a displayed math environment. LaTeX catches the mistake, but the line number reported in the error message is incorrect.

**Example 6** Add a $ somewhere in line 53 (such errors often occur when copying and pasting formulas):

```
\langle $\ldots, 0, \ldots, d, \ldots, 0, \ldots \rangle
```

This results in the message:

```
! Display math should end with $$.
<to be read again>
                   \protect
l.53       \langle $\ldots
                         , 0, \ldots, d, \ldots, 0, \ldots \rangle
```

You cannot have a $ in a displayed math formula.

Error messages from LaTeX are not always as helpful as they could be, but there is always some information that can be gleaned from them. As a rule, the error message should at least tell you the line number (or paragraph or formula) where LaTeX realizes that there is a problem. Try to identify the structure (the command or environment) that caused the error. Keep in mind that the error could be quite far from the line LaTeX indicates but will always be on or before that line in the source file.

If you have difficulty isolating a problem, create a `current.tex` file that has the same preamble as your current source file and an empty `document` environment. Then copy the paragraphs you suspect might have problems into this document one by one and track down the errors. Once your new document typesets correctly, copy the paragraph back into your real document, and work on another paragraph. If necessary, split a large paragraph into smaller pieces.

Finally, typeset often. Typesetting this book with the closing brace of the first `\caption` removed (from the figure on page xiii) gives the error message

```
! Text line contains an invalid character.
l.1557 ...pletely irreducible^^?
```

where the reference (`l.1557`) is to the text in the middle of page 3. However, if the only thing you did before typesetting was to insert a figure, at least you know where to look for errors. If you make a dozen corrections and then typeset, you may not know where to begin to look.

## 6.2 *Logical and visual design*

The goal of this book is to teach you how to *typeset* an article; not how to *write* it. The typeset version of `intrart.tex` (pp. 53–54) looks impressive. The typeset version of `sampart.tex` (pp. 67–69) even more so. To produce such articles, you need to understand that there are two aspects to article design: the *visual* and the *logical*.

As an example, let us look at a theorem from `intrart.tex`, the LaTeX sample article (the typeset form of the theorem is on page 54). You tell LaTeX that you want to state a theorem by using a `theorem` environment:

```
\begin{theorem}\label{T:P*}
   Let $D_{i}$, $i \in I$, be complete distributive
   lattices satisfying condition~\textup{(J)}.  Let $\Theta$
   be a complete congruence relation on
   $\Pi^{*} ( D_{i} \mid i \in I )$.
   If there exist $i \in I$ and $d \in D_{i}$ with
   $d < 1_{i}$ such that, for all $d \leq c < 1_{i}$,
   \begin{equation} \label{E:cong1}
      \langle \ldots, d, \ldots, 0, \ldots \rangle \equiv
      \langle \ldots, c, \ldots, 0, \ldots \rangle \pmod{\Theta},
   \end{equation}
```

```
    then $\Theta = \iota$.
\end{theorem}
```

The logical part of the design is choosing to define a theorem by placing material inside a theorem environment. For the visual design, LaTeX must make hundreds of decisions. Could you specify all of the spacing, font size changes, centering, numbering, and so on? Maybe, but would you *want* to? And would you want to repeat that process for every theorem in your document?

Even if you did, you would have spent a great deal of time and energy on the ***visual design*** of the theorems, rather than on the ***logical design*** of your article. The idea behind LaTeX is that you should concentrate on what you have to say and let LaTeX take care of the visual design.

LaTeX uses four major tools to separate the logical and visual design of an article:

1. **Commands** Information is given to LaTeX in the arguments of commands; that information is then typeset according to the definitions of the commands. In the `sampart.tex` sample article (see the `samples` directory and pages 67–77), the construct $D^{\langle 2 \rangle}$ is often used. We can define

   ```
   \newcommand{\Ds}{D^{\langle 2 \rangle}}
   ```

   and use `\Ds` in place of `D^{\langle 2 \rangle}`. If a referee or co-author later suggests different notation, editing this *one line* will change the notation throughout the whole article.

2. **Environments** Important logical structures are placed within environments. For example, list items are typed within a list environment and formatted accordingly. If you later decide to change the type of the list, you can do so by simply changing the name of the environment.

3. **Proclamations** You can change the style or the numbering scheme of any proclamation at any time by changing that proclamation's definition in the preamble, especially if you use the `amsart` document class. 

4. **Numbering and cross-referencing** Theorems, lemmas, definitions, and sections are logical units that can be freely moved around. LaTeX recalculates the numbering and the cross-references.

You write articles to communicate your ideas. The closer you get to a separation of logical and visual design, the more you can concentrate on that goal.

## 6.3 How LaTeX works

Now that you have learned how to use LaTeX to typeset an article, it is time to get a brief overview of how LaTeX works. As mentioned in the introduction, LaTeX's core is a programming language called TeX created by Donald Knuth, which provides low-level typesetting instructions. TeX comes with a set of fonts called *Computer Modern* (CM). The CM fonts and the TeX programming language form the foundation of a typical TeX system.

TeX is expandable—new commands can be defined in terms of more basic ones. LaTeX is one of the best known expansions of TeX, introducing the concept of *logical units,* which you read about in Section 6.2, and adding a large number of higher-level commands.

Ⓐ The visual layout of LaTeX documents is primarily determined by the ***document class*** (you now have some familiarity with two document classes, `article` and `amsart`; other standard classes include `book`, `letter`, `report`, and `slides`). Many journals, publishers, and schools have their own document classes for formatting articles, books, and theses.

Ⓐ Expansions of LaTeX are called *packages* (we have already come across a number of these, including the packages amsmath, amssymb, amstext, amsthm, eufrak, and latexsym); they add new functionality to LaTeX (by adding new commands and environments) or change the way previously defined commands and environments work. It is essential that you find the packages that make your work easier. (See Section C.1.)

My view of the structure of TeX and LaTeX is illustrated in Figure 6.1. This figure suggests that in order to work with a LaTeX document, you first have to install TeX and the CM fonts, then LaTeX, and finally specify the document class and

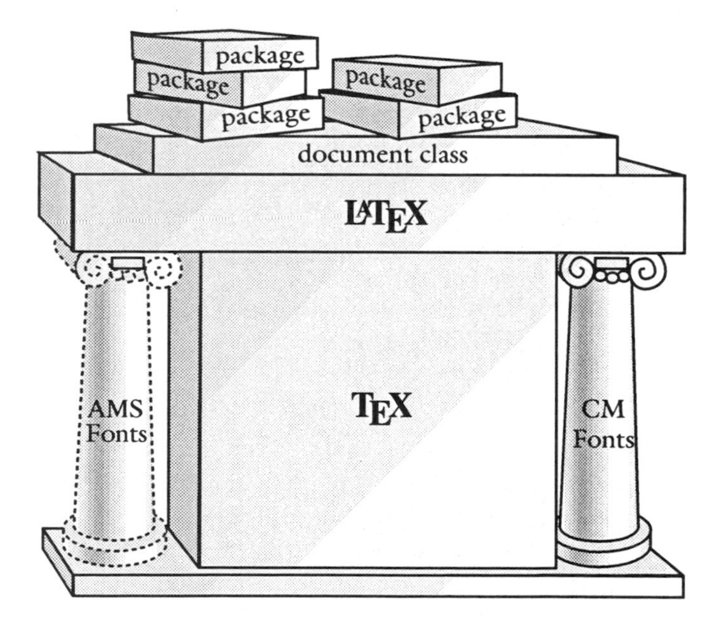


Figure 6.1: The structure of LaTeX.

the necessary packages. The AMSFonts font set is very useful but not absolutely necessary.

## 6.4 Using LaTeX

Figure 6.2 illustrates the steps in the production of a typeset document. We start by opening an existing source file or creating a new one with a text editor (for this discussion, the LaTeX source file will be called `myart.tex`). Once the source file is ready, we typeset it using the `LaTeX` format. We will end up with at least three files:

1. `myart.dvi` The typeset article in machine-readable format (DVI stands for device independent).
2. `myart.aux` The auxiliary file, used by LaTeX for internal "bookkeeping," including cross-referencing and bibliographic citations.
3. `myart.log` The log file, where LaTeX records the messages generated during the typesetting session, including the warnings and error messages.

The computer uses a *video driver* (DVI *viewer*) to display the typeset article, `myart.dvi`, on the monitor; a *printer driver* to print the typeset article on a printer; and a *PostScript driver* (DVI to PostScript *converter*) to convert the typeset article to PostScript format. (For Macintosh and PC implementations of TeX, the PostScript converters are often in the "Save as" option of the printer driver dialog box; for most UNIX implementations, the printer driver and the PostScript driver are separate applications.)

It should be emphasized that of the four programs used (TeX and the three drivers), only one (TeX) is the same for all computers and all implementations. If you use TeX in an "integrated environment," the four programs appear as one.

### 6.4.1 LaTeX prompts

If LaTeX cannot carry out your instructions, it displays a *prompt* (and possibly an error message; see Section 6.1):

- The `**` prompt means that LaTeX needs to know the name of a source file to typeset. This usually means that you have misspelled a file name, you are trying to typeset a document that is not located in TeX's current directory, or that there is a space in the file name of your source file.
- The `?` prompt indicates that LaTeX has found an error in your source file, and wants you to decide what to do next. You can try to continue typesetting the file by pressing Return. Depending on the nature of the error, LaTeX may either

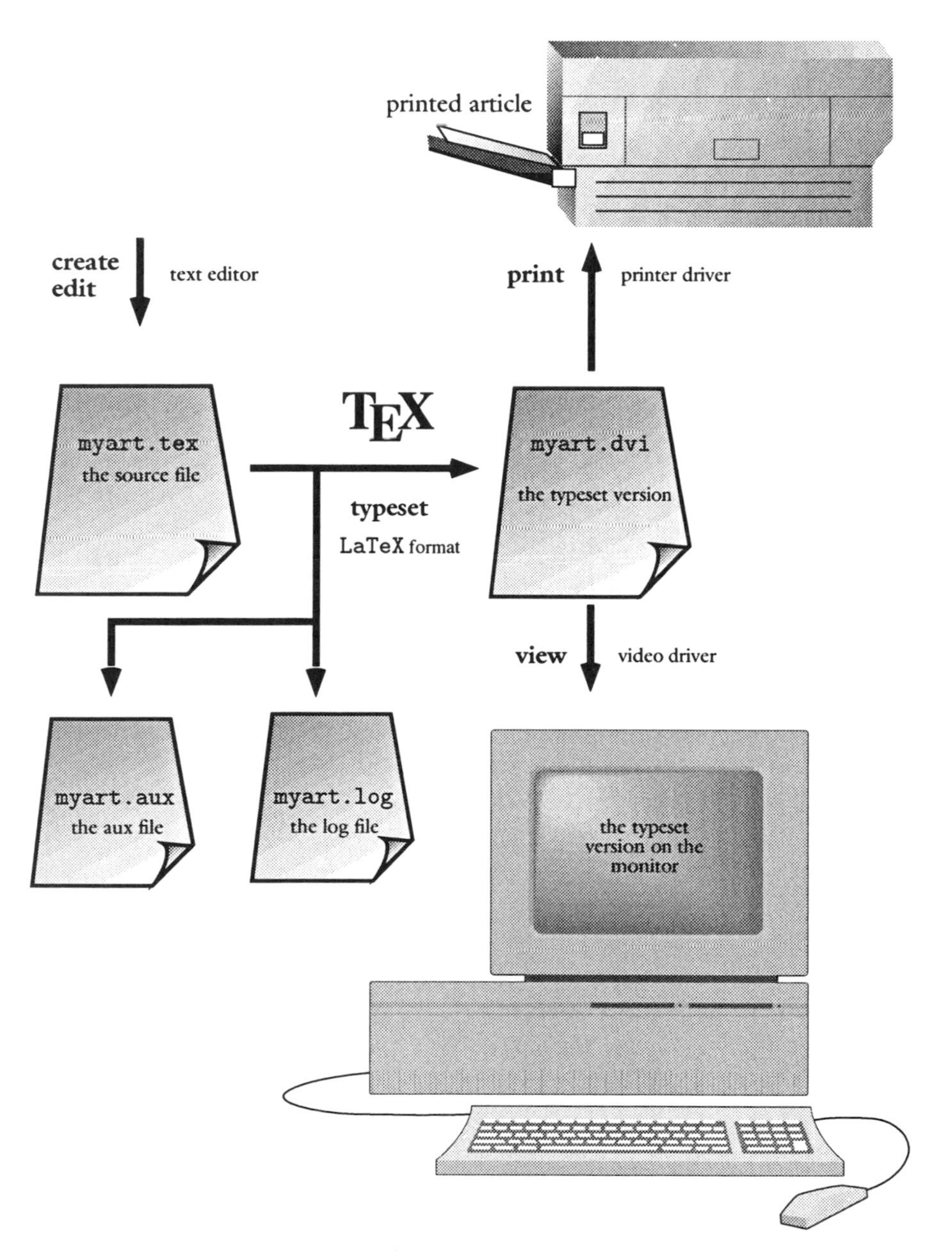


Figure 6.2: Using LATEX.

recover or generate additional error messages. You can also type x to stop typesetting your file, or h, which may give you some useful advice on how to correct the error.

- If you misspelled the name of a package in a \usepackage command, or if LaTeX cannot find a file, it will display a message such as the following:

```
! LaTeX Error: File 'misspelled.sty' not found.

Type X to quit or <RETURN> to proceed,
or enter new name. (Default extension: sty)

Enter file name:
```

  You can either type the correct name of the file, or type x to quit LaTeX.

- The * prompt signifies that LaTeX is in *interactive mode* and is waiting for instructions. To get to such a prompt, comment out the line

```
\end{document}
```

  in a source file by inserting a % symbol as the first character of the line, then typeset the file. To exit interactive mode, type

```
\end{document}
```

  at the * prompt, and press Return.

In interactive mode you can type in a fragment from your source file to see how it will be typeset, or use a command such as \show to get the definition of a command. For example, if you type

```
*\show \vec
```

LaTeX responds with

```
> \vec=macro:
->\mathaccent "017E
<*> \show \vec
```

which tells you that \vec is a command that produces a math accent.

### 6.4.2 Versions

A complete LaTeX distribution consists of hundreds of files, all of which interact. Since these files have had many revisions, you should make sure that they are all up-to-date and compatible with each other. You can check the version numbers and dates by reading the first few lines of each file in a text editor or by checking the dates and version numbers that are shown on the list created by the `\listfiles` command, which will be discussed later in this section.

LaTeX is updated every six months; while writing this book, the version issued on June 30, 1998 was used.

When you typeset a LaTeX document, LaTeX prints its release date in the log file with a line such as

```
LaTeX2e <1998/06/30>
```

If you use a LaTeX feature that was introduced recently, you can put a command such as the following into the preamble of your source file:

```
\NeedsTeXFormat{LaTeX2e}[1998/12/01]
```

This command specifies the date of the oldest version of LaTeX that may be used to typeset your file. If someone attempts to typeset your file with an older version, LaTeX will generate a warning.

As of this writing, the AMS packages ($\mathcal{A}\mathcal{M}\mathcal{S}$-LaTeX) are at version 1.2 (make sure they are no older than October 24, 1996) and the AMSFonts set is at version 2.2. See Section C.1 for more information on obtaining updated versions.

If you include the `\listfiles` command in the preamble of your document, the log file will contain a detailed listing of all the files used in the typesetting.

Here are a few (truncated) lines from such a listing:

```
 *File List*
       book.cls   1998/05/05 v1.3y Standard LaTeX document class
      leqno.clo   1996/07/26 v1.1b Standard LaTeX option
       bk10.clo   1998/05/05 v1.3y Standard LaTeX file
FirstSteps.sty   1999/03/15 Commands for First Steps book
    amsmath.sty   1997/03/20 v1.2d AMS math features
    amstext.sty   1996/10/28 v1.2b
   latexsym.sty   1996/11/20 v2.2d Standard LaTeX package
   graphics.sty   1997/09/09 v1.0f Standard LaTeX Graphics
```

### 6.4.3 *Spell checkers and text editors*

It can be very frustrating to try to check the spelling of a LaTeX document with a regular spell checker because it will try to spell check your math! Luckily, there are LaTeX-aware spell-checking applications available for all three major platforms:

- Macintosh
  - Excalibur
    `http://www.eg.bucknell.edu/~excalibr/excalibur.html`

- PC
  - jspell
    `ftp://ftp.tex.ac.uk/pub/archive/support/jspell/`
  - Trigram Systems' Microspell
    Available from Y&Y, among other sources.
  - TeXSpell
    Comes with PCTeX for Windows.
- UNIX
  - ispell is the spell checker of choice.

All three computer platforms also have a variety of freeware, shareware, and commercial "TeX-aware" text editors with features including syntax coloring, automatic completion of LaTeX commands and environments, and management of `\label` and `\ref` entries. Using such a text editor can make working with LaTeX much more pleasant and efficient.

### 6.4.4 LaTeX for non-English languages

LaTeX was originally designed to work with (American) English text, but during the last decade great progress has been made in extending LaTeX to work with other languages.

The babel package by Johannes Braams (part of the standard LaTeX distribution) provides the foundation for this support. You will also need the hyphenation documents for the languages you want to work with, and the Extended Computer Modern or the European Modern (EC or EM) fonts.

There are two more important packages:

- The fontenc package allows experts to specify the appropriate encoding for the language.
- With the inputenc package, you can enter composite characters, such as accented characters appropriate to the language—into the LaTeX source file.

Consult your local TeX user group to find out what is available and how to proceed (see Section C.4).

## 6.5 Further reading

Naturally, I hope that you will read the "big brother" of this book, *Math into LaTeX,* third edition, published by Birkhäuser, Boston, and Springer-Verlag, New York, 1999, for a more complete discussion of LaTeX. Reference [1], below, is strongly recommended for learning how to modify LaTeX to suit your needs and for an overview of the most important LaTeX packages; reference [2] on PostScript illustrations in LaTeX documents; and reference [3] on LaTeX and the World Wide Web. The sale of books [1]–[3] also helps support the LaTeX3 team.

[1] Michel Goossens, Frank Mittelbach, and Alexander Samarin,
*The LaTeX Companion.*
Addison-Wesley, Reading, MA. 1994.

[2] Michel Goossens, Sebastian Rahtz (Contributor), and Frank Mittelbach,
*The LaTeX Graphics Companion.*
Addison-Wesley, Reading, MA. 1997.

[3] Michel Goossens and Sebastian Rahtz, with Eitan Gurari, Ross Moore, and Robert Sutor,
*The LaTeX Web Companion: Integrating TeX, HTML and XML.*
Addison-Wesley, Reading, MA. 1999.

[4] Donald E. Knuth,
*The TeXbook.* Computers and Typesetting. Vol. A.
Addison-Wesley, Reading, MA. 1984, 1990.

[5] Leslie Lamport,
*LaTeX: A Document Preparation System,* second edition.
Addison-Wesley, Reading, MA. 1994.

APPENDIX

# Math symbol tables

## A.1 Hebrew and Greek letters

### A.1.1 Hebrew letters

| Source | Type | Print |
|---|---|---|
| LaTeX | \aleph | ℵ |
| amssymb | \beth | ℶ |
| | \daleth | ℸ |
| | \gimel | ℷ |

## *A.1.2 Greek letters*

| Type | Print | Type | Print | Type | Print |
|---|---|---|---|---|---|
| \alpha | $\alpha$ | \vartheta | $\vartheta$ | \rho | $\rho$ |
| \beta | $\beta$ | \iota | $\iota$ | \varrho | $\varrho$ |
| \gamma | $\gamma$ | \kappa | $\kappa$ | \sigma | $\sigma$ |
| \digamma | $\digamma$ | \varkappa | $\varkappa$ | \varsigma | $\varsigma$ |
| \delta | $\delta$ | \lambda | $\lambda$ | \tau | $\tau$ |
| \epsilon | $\epsilon$ | \mu | $\mu$ | \upsilon | $\upsilon$ |
| \varepsilon | $\varepsilon$ | \nu | $\nu$ | \phi | $\phi$ |
| \zeta | $\zeta$ | \xi | $\xi$ | \varphi | $\varphi$ |
| \eta | $\eta$ | \pi | $\pi$ | \chi | $\chi$ |
| \theta | $\theta$ | \varpi | $\varpi$ | \psi | $\psi$ |
| | | | | \omega | $\omega$ |

\digamma and \varkappa require the amssymb package. Ⓐ

| Type | Print | Type | Print |
|---|---|---|---|
| \Gamma | $\Gamma$ | \varGamma | $\varGamma$ |
| \Delta | $\Delta$ | \varDelta | $\varDelta$ |
| \Theta | $\Theta$ | \varTheta | $\varTheta$ |
| \Lambda | $\Lambda$ | \varLambda | $\varLambda$ |
| \Xi | $\Xi$ | \varXi | $\varXi$ |
| \Pi | $\Pi$ | \varPi | $\varPi$ |
| \Sigma | $\Sigma$ | \varSigma | $\varSigma$ |
| \Upsilon | $\Upsilon$ | \varUpsilon | $\varUpsilon$ |
| \Phi | $\Phi$ | \varPhi | $\varPhi$ |
| \Psi | $\Psi$ | \varPsi | $\varPsi$ |
| \Omega | $\Omega$ | \varOmega | $\varOmega$ |

All symbols whose name begins with var require the amsmath package. Ⓐ

# A.2 Binary relations

## A.2.1 LaTeX binary relations

| Source | Type | Print | Type | Print |
|---|---|---|---|---|
| LaTeX | \in | $\in$ | \ni | $\ni$ |
| | \leq | $\leq$ | \geq | $\geq$ |
| | \ll | $\ll$ | \gg | $\gg$ |
| | \prec | $\prec$ | \succ | $\succ$ |
| | \preceq | $\preceq$ | \succeq | $\succeq$ |
| | \sim | $\sim$ | \approx | $\approx$ |
| | \simeq | $\simeq$ | \cong | $\cong$ |
| | \equiv | $\equiv$ | \doteq | $\doteq$ |
| | \subset | $\subset$ | \supset | $\supset$ |
| | \subseteq | $\subseteq$ | \supseteq | $\supseteq$ |
| | \sqsubseteq | $\sqsubseteq$ | \sqsupseteq | $\sqsupseteq$ |
| | \smile | $\smile$ | \frown | $\frown$ |
| | \perp | $\perp$ | \models | $\models$ |
| | \mid | $\mid$ | \parallel | $\parallel$ |
| | \vdash | $\vdash$ | \dashv | $\dashv$ |
| | \propto | $\propto$ | \asymp | $\asymp$ |
| | \bowtie | $\bowtie$ | | |
| latexsym | \sqsubset | $\sqsubset$ | \sqsupset | $\sqsupset$ |
| | \Join | $\Join$ | | |

## *A.2.2 AMS binary relations*

| Type | Print | Type | Print |
|---|---|---|---|
| \leqslant | $\leqslant$ | \geqslant | $\geqslant$ |
| \eqslantless | $\eqslantless$ | \eqslantgtr | $\eqslantgtr$ |
| \lesssim | $\lesssim$ | \gtrsim | $\gtrsim$ |
| \lessapprox | $\lessapprox$ | \gtrapprox | $\gtrapprox$ |
| \approxeq | $\approxeq$ | | |
| \lessdot | $\lessdot$ | \gtrdot | $\gtrdot$ |
| \lll | $\lll$ | \ggg | $\ggg$ |
| \lessgtr | $\lessgtr$ | \gtrless | $\gtrless$ |
| \lesseqgtr | $\lesseqgtr$ | \gtreqless | $\gtreqless$ |
| \lesseqqgtr | $\lesseqqgtr$ | \gtreqqless | $\gtreqqless$ |
| \doteqdot | $\doteqdot$ | \eqcirc | $\eqcirc$ |
| \circeq | $\circeq$ | \triangleq | $\triangleq$ |
| \risingdotseq | $\risingdotseq$ | \fallingdotseq | $\fallingdotseq$ |
| \backsim | $\backsim$ | \thicksim | $\thicksim$ |
| \backsimeq | $\backsimeq$ | \thickapprox | $\thickapprox$ |
| \preccurlyeq | $\preccurlyeq$ | \succcurlyeq | $\succcurlyeq$ |
| \curlyeqprec | $\curlyeqprec$ | \curlyeqsucc | $\curlyeqsucc$ |
| \precsim | $\precsim$ | \succsim | $\succsim$ |
| \precapprox | $\precapprox$ | \succapprox | $\succapprox$ |
| \subseteqq | $\subseteqq$ | \supseteqq | $\supseteqq$ |
| \Subset | $\Subset$ | \Supset | $\Supset$ |
| \vartriangleleft | $\vartriangleleft$ | \vartriangleright | $\vartriangleright$ |
| \trianglelefteq | $\trianglelefteq$ | \trianglerighteq | $\trianglerighteq$ |
| \vDash | $\vDash$ | \Vdash | $\Vdash$ |
| \Vvdash | $\Vvdash$ | | |
| \smallsmile | $\smallsmile$ | \smallfrown | $\smallfrown$ |
| \shortmid | $\shortmid$ | \shortparallel | $\shortparallel$ |
| \bumpeq | $\bumpeq$ | \Bumpeq | $\Bumpeq$ |
| \between | $\between$ | \pitchfork | $\pitchfork$ |
| \varpropto | $\varpropto$ | \backepsilon | $\backepsilon$ |
| \blacktriangleleft | $\blacktriangleleft$ | \blacktriangleright | $\blacktriangleright$ |
| \therefore | $\therefore$ | \because | $\because$ |

All symbols require the amssymb package. Ⓐ

### A.2.3 AMS negated binary relations

| Type | Print | Type | Print |
|---|---|---|---|
| \ne | $\ne$ | \notin | $\notin$ |
| \nless | $\nless$ | \ngtr | $\ngtr$ |
| \nleq | $\nleq$ | \ngeq | $\ngeq$ |
| \nleqslant | $\nleqslant$ | \ngeqslant | $\ngeqslant$ |
| \nleqq | $\nleqq$ | \ngeqq | $\ngeqq$ |
| \lneq | $\lneq$ | \gneq | $\gneq$ |
| \lneqq | $\lneqq$ | \gneqq | $\gneqq$ |
| \lvertneqq | $\lvertneqq$ | \gvertneqq | $\gvertneqq$ |
| \lnsim | $\lnsim$ | \gnsim | $\gnsim$ |
| \lnapprox | $\lnapprox$ | \gnapprox | $\gnapprox$ |
| \nprec | $\nprec$ | \nsucc | $\nsucc$ |
| \npreceq | $\npreceq$ | \nsucceq | $\nsucceq$ |
| \precneqq | $\precneqq$ | \succneqq | $\succneqq$ |
| \precnsim | $\precnsim$ | \succnsim | $\succnsim$ |
| \precnapprox | $\precnapprox$ | \succnapprox | $\succnapprox$ |
| \nsim | $\nsim$ | \ncong | $\ncong$ |
| \nshortmid | $\nshortmid$ | \nshortparallel | $\nshortparallel$ |
| \nmid | $\nmid$ | \nparallel | $\nparallel$ |
| \nvdash | $\nvdash$ | \nvDash | $\nvDash$ |
| \nVdash | $\nVdash$ | \nVDash | $\nVDash$ |
| \ntriangleleft | $\ntriangleleft$ | \ntriangleright | $\ntriangleright$ |
| \ntrianglelefteq | $\ntrianglelefteq$ | \ntrianglerighteq | $\ntrianglerighteq$ |
| \nsubseteq | $\nsubseteq$ | \nsupseteq | $\nsupseteq$ |
| \nsubseteqq | $\nsubseteqq$ | \nsupseteqq | $\nsupseteqq$ |
| \subsetneq | $\subsetneq$ | \supsetneq | $\supsetneq$ |
| \varsubsetneq | $\varsubsetneq$ | \varsupsetneq | $\varsupsetneq$ |
| \subsetneqq | $\subsetneqq$ | \supsetneqq | $\supsetneqq$ |
| \varsubsetneqq | $\varsubsetneqq$ | \varsupsetneqq | $\varsupsetneqq$ |

Ⓐ All symbols but \ne require the amssymb package.

# *A.3 Binary operations*

| Source | Type | Print | Type | Print |
|---|---|---|---|---|
| LaTeX | \pm | $\pm$ | \mp | $\mp$ |
| | \times | $\times$ | \cdot | $\cdot$ |
| | \circ | $\circ$ | \bigcirc | $\bigcirc$ |
| | \div | $\div$ | \diamond | $\diamond$ |
| | \ast | $\ast$ | \star | $\star$ |
| | \cap | $\cap$ | \cup | $\cup$ |
| | \sqcap | $\sqcap$ | \sqcup | $\sqcup$ |
| | \wedge | $\wedge$ | \vee | $\vee$ |
| | \triangleleft | $\triangleleft$ | \triangleright | $\triangleright$ |
| | \bigtriangleup | $\bigtriangleup$ | \bigtriangledown | $\bigtriangledown$ |
| | \oplus | $\oplus$ | \ominus | $\ominus$ |
| | \otimes | $\otimes$ | \oslash | $\oslash$ |
| | \odot | $\odot$ | \bullet | $\bullet$ |
| | \dagger | $\dagger$ | \ddagger | $\ddagger$ |
| | \setminus | $\setminus$ | \uplus | $\uplus$ |
| | \wr | $\wr$ | \amalg | $\amalg$ |
| latexsym | \lhd | $\lhd$ | \rhd | $\rhd$ |
| | \unlhd | $\unlhd$ | \unrhd | $\unrhd$ |
| amssymb | \dotplus | $\dotplus$ | \centerdot | $\centerdot$ |
| | \ltimes | $\ltimes$ | \rtimes | $\rtimes$ |
| | \leftthreetimes | $\leftthreetimes$ | \rightthreetimes | $\rightthreetimes$ |
| | \circleddash | $\circleddash$ | \smallsetminus | $\smallsetminus$ |
| | \barwedge | $\barwedge$ | \doublebarwedge | $\doublebarwedge$ |
| | \curlywedge | $\curlywedge$ | \curlyvee | $\curlyvee$ |
| | \veebar | $\veebar$ | \intercal | $\intercal$ |
| | \Cap | $\Cap$ | \Cup | $\Cup$ |
| | \circledast | $\circledast$ | \circledcirc | $\circledcirc$ |
| | \boxminus | $\boxminus$ | \boxtimes | $\boxtimes$ |
| | \boxdot | $\boxdot$ | \boxplus | $\boxplus$ |
| | \divideontimes | $\divideontimes$ | \vartriangle | $\vartriangle$ |
| amsmath | \And | & | | |

# A.4 Arrows

| Source | Type | Print | Type | Print |
|---|---|---|---|---|
| LaTeX | `\leftarrow` | $\leftarrow$ | `\rightarrow` or `\to` | $\rightarrow$ |
| | `\longleftarrow` | $\longleftarrow$ | `\longrightarrow` | $\longrightarrow$ |
| | `\Leftarrow` | $\Leftarrow$ | `\Rightarrow` | $\Rightarrow$ |
| | `\Longleftarrow` | $\Longleftarrow$ | `\Longrightarrow` | $\Longrightarrow$ |
| | `\leftrightarrow` | $\leftrightarrow$ | `\longleftrightarrow` | $\longleftrightarrow$ |
| | `\Leftrightarrow` | $\Leftrightarrow$ | `\Longleftrightarrow` | $\Longleftrightarrow$ |
| | `\uparrow` | $\uparrow$ | `\downarrow` | $\downarrow$ |
| | `\Uparrow` | $\Uparrow$ | `\Downarrow` | $\Downarrow$ |
| | `\updownarrow` | $\updownarrow$ | `\Updownarrow` | $\Updownarrow$ |
| | `\nearrow` | $\nearrow$ | `\searrow` | $\searrow$ |
| | `\swarrow` | $\swarrow$ | `\nwarrow` | $\nwarrow$ |
| | `\mapsto` | $\mapsto$ | `\longmapsto` | $\longmapsto$ |
| | `\hookleftarrow` | $\hookleftarrow$ | `\hookrightarrow` | $\hookrightarrow$ |
| | `\leftharpoonup` | $\leftharpoonup$ | `\rightharpoonup` | $\rightharpoonup$ |
| | `\leftharpoondown` | $\leftharpoondown$ | `\rightharpoondown` | $\rightharpoondown$ |
| | `\leftrightharpoons` | $\leftrightharpoons$ | `\rightleftharpoons` | $\rightleftharpoons$ |
| latexsym | `\leadsto` | $\leadsto$ | | |
| amssymb | `\leftleftarrows` | $\leftleftarrows$ | `\rightrightarrows` | $\rightrightarrows$ |
| | `\leftrightarrows` | $\leftrightarrows$ | `\rightleftarrows` | $\rightleftarrows$ |
| | `\Lleftarrow` | $\Lleftarrow$ | `\Rrightarrow` | $\Rrightarrow$ |
| | `\twoheadleftarrow` | $\twoheadleftarrow$ | `\twoheadrightarrow` | $\twoheadrightarrow$ |
| | `\leftarrowtail` | $\leftarrowtail$ | `\rightarrowtail` | $\rightarrowtail$ |
| | `\looparrowleft` | $\looparrowleft$ | `\looparrowright` | $\looparrowright$ |
| | `\upuparrows` | $\upuparrows$ | `\downdownarrows` | $\downdownarrows$ |
| | `\upharpoonleft` | $\upharpoonleft$ | `\upharpoonright` | $\upharpoonright$ |
| | `\downharpoonleft` | $\downharpoonleft$ | `\downharpoonright` | $\downharpoonright$ |
| | `\leftrightsquigarrow` | $\leftrightsquigarrow$ | `\rightsquigarrow` | $\rightsquigarrow$ |
| | `\multimap` | $\multimap$ | | |
| amssymb | `\nleftarrow` | $\nleftarrow$ | `\nrightarrow` | $\nrightarrow$ |
| | `\nLeftarrow` | $\nLeftarrow$ | `\nRightarrow` | $\nRightarrow$ |
| | `\nleftrightarrow` | $\nleftrightarrow$ | `\nLeftrightarrow` | $\nLeftrightarrow$ |

# *A.5 Miscellaneous symbols*

| Source | Type | Print | Type | Print |
|---|---|---|---|---|
| LaTeX | \hbar | $\hbar$ | \ell | $\ell$ |
| | \imath | $\imath$ | \jmath | $\jmath$ |
| | \wp | $\wp$ | \Re | $\Re$ |
| | \Im | $\Im$ | \partial | $\partial$ |
| | \infty | $\infty$ | \prime | $\prime$ |
| | \emptyset | $\emptyset$ | \neg | $\neg$ |
| | \forall | $\forall$ | \exists | $\exists$ |
| | \smallint | $\smallint$ | \triangle | $\triangle$ |
| | \top | $\top$ | \bot | $\bot$ |
| | \P | ¶ | \S | § |
| | \dag | † | \ddag | ‡ |
| | \flat | $\flat$ | \natural | $\natural$ |
| | \sharp | $\sharp$ | \angle | $\angle$ |
| | \clubsuit | $\clubsuit$ | \diamondsuit | $\diamondsuit$ |
| | \heartsuit | $\heartsuit$ | \spadesuit | $\spadesuit$ |
| | \surd | $\surd$ | | |
| latexsym | \Box | $\Box$ | \Diamond | $\Diamond$ |
| | \mho | $\mho$ | | |
| amssymb | \hslash | $\hslash$ | \complement | $\complement$ |
| | \backprime | $\backprime$ | \nexists | $\nexists$ |
| | \Bbbk | $\Bbbk$ | \varnothing | $\varnothing$ |
| | \diagup | $\diagup$ | \diagdown | $\diagdown$ |
| | \blacktriangle | $\blacktriangle$ | \blacktriangledown | $\blacktriangledown$ |
| | \triangledown | $\triangledown$ | \Game | $\Game$ |
| | \square | $\square$ | \blacksquare | $\blacksquare$ |
| | \lozenge | $\lozenge$ | \blacklozenge | $\blacklozenge$ |
| | \measuredangle | $\measuredangle$ | \sphericalangle | $\sphericalangle$ |
| | \circledS | $\circledS$ | \bigstar | $\bigstar$ |
| | \Finv | $\Finv$ | \eth | $\eth$ |

## A.6 Math and text spacing commands

| Short form | Full form | Size | Short form | Full form |
|---|---|---|---|---|
| `\,` | `\thinspace` | ␣ | `\!` | `\negthinspace` |
| `\:` | `\medspace` | ␣ | | `\negmedspace` |
| `\;` | `\thickspace` | ␣ | | `\negthickspace` |
| | `\quad` | ␣ | | |
| | `\qquad` | ␣ | | |

Ⓐ The `\medspace`, `\thickspace`, `\negmedspace`, and `\negthickspace` commands require the amsmath package. Two additional spacing commands, the interword space (`\␣`) and the tie (`~`), do not have full forms and are not included in the table.

## A.7 Delimiters

<table>
<tr><th>Name</th><th>Type</th><th>Print</th><th>Name</th><th>Type</th><th>Print</th></tr>
<tr><td>Left parenthesis</td><td><code>(</code></td><td>(</td><td>Right parenthesis</td><td><code>)</code></td><td>)</td></tr>
<tr><td>Left bracket</td><td><code>[</code></td><td>[</td><td>Right bracket</td><td><code>]</code></td><td>]</td></tr>
<tr><td>Left brace</td><td><code>\{</code></td><td>{</td><td>Right brace</td><td><code>\}</code></td><td>}</td></tr>
<tr><td>Reverse slash</td><td><code>\backslash</code></td><td>\</td><td>Forward slash</td><td><code>/</code></td><td>/</td></tr>
<tr><td>Left angle</td><td><code>\langle</code></td><td>⟨</td><td>Right angle</td><td><code>\rangle</code></td><td>⟩</td></tr>
<tr><td>Vertical line</td><td><code>|</code> or <code>\vert</code></td><td>|</td><td>Double vert. line</td><td><code>\|</code> or <code>\Vert</code></td><td>‖</td></tr>
<tr><td>Left floor</td><td><code>\lfloor</code></td><td>⌊</td><td>Right floor</td><td><code>\rfloor</code></td><td>⌋</td></tr>
<tr><td>Left ceiling</td><td><code>\lceil</code></td><td>⌈</td><td>Right ceiling</td><td><code>\rceil</code></td><td>⌉</td></tr>
<tr><td>Upper left corner</td><td><code>\ulcorner</code></td><td>⌜</td><td>Upper right corner</td><td><code>\urcorner</code></td><td>⌝</td></tr>
<tr><td>Lower left corner</td><td><code>\llcorner</code></td><td>⌞</td><td>Lower right corner</td><td><code>\lrcorner</code></td><td>⌟</td></tr>
</table>

Ⓐ The corners require the amsmath package.

| Name | Type | Print |
|---|---|---|
| Upward arrow | `\uparrow` | ↑ |
| Double upward arrow | `\Uparrow` | ⇑ |
| Downward arrow | `\downarrow` | ↓ |
| Double downward arrow | `\Downarrow` | ⇓ |
| Up-and-down arrow | `\updownarrow` | ↕ |
| Double up-and-down arrow | `\Updownarrow` | ⇕ |

# *A.8 Operators*

| \arccos | \cos | \csc | \ker | \sec |
|---|---|---|---|---|
| \arcsin | \cosh | \dim | \lg | \sin |
| \arctan | \cot | \exp | \ln | \sinh |
| \arg | \coth | \hom | \log | \tan |
| | | | | \tanh |

"Pure" operators, no limit.

| \det (det) | \min (min) | \injlim (inj lim) | \varinjlim ($\varinjlim$) |
|---|---|---|---|
| \gcd (gcd) | \max (max) | \liminf (lim inf) | \varliminf ($\varliminf$) |
| \inf (inf) | \Pr (Pr) | \limsup (lim sup) | \varlimsup ($\varlimsup$) |
| \lim (lim) | \sup (sup) | \projlim (proj lim) | \varprojlim ($\varprojlim$) |

The operators in this table can have a "limit," which is typed as a subscript. The \var commands, \injlim, and \projlim require the amsmath package. 

### *A.8.1 Large operators*

| Type | Inline | Displayed | Type | Inline | Displayed |
|---|---|---|---|---|---|
| \int_{a}^{b} | $\int_a^b$ | $\displaystyle\int_a^b$ | \oint_{a}^{b} | $\oint_a^b$ | $\displaystyle\oint_a^b$ |
| \prod_{i=1}^{n} | $\prod_{i=1}^n$ | $\displaystyle\prod_{i=1}^n$ | \coprod_{i=1}^{n} | $\coprod_{i=1}^n$ | $\displaystyle\coprod_{i=1}^n$ |
| \bigcap_{i=1}^{n} | $\bigcap_{i=1}^n$ | $\displaystyle\bigcap_{i=1}^n$ | \bigcup_{i=1}^{n} | $\bigcup_{i=1}^n$ | $\displaystyle\bigcup_{i=1}^n$ |
| \bigwedge_{i=1}^{n} | $\bigwedge_{i=1}^n$ | $\displaystyle\bigwedge_{i=1}^n$ | \bigvee_{i=1}^{n} | $\bigvee_{i=1}^n$ | $\displaystyle\bigvee_{i=1}^n$ |
| \bigsqcup_{i=1}^{n} | $\bigsqcup_{i=1}^n$ | $\displaystyle\bigsqcup_{i=1}^n$ | \biguplus_{i=1}^{n} | $\biguplus_{i=1}^n$ | $\displaystyle\biguplus_{i=1}^n$ |
| \bigotimes_{i=1}^{n} | $\bigotimes_{i=1}^n$ | $\displaystyle\bigotimes_{i=1}^n$ | \bigoplus_{i=1}^{n} | $\bigoplus_{i=1}^n$ | $\displaystyle\bigoplus_{i=1}^n$ |
| \bigodot_{i=1}^{n} | $\bigodot_{i=1}^n$ | $\displaystyle\bigodot_{i=1}^n$ | \sum_{i=1}^{n} | $\sum_{i=1}^n$ | $\displaystyle\sum_{i=1}^n$ |

# A.9 Math accents and fonts

## A.9.1 Math accents

| | | | | | | | |
|---|---|---|---|---|---|---|---|
| `\hat{a}` | $\hat{a}$ | `\Hat{a}` | $\Hat{a}$ | `\widehat{a}` | $\widehat{a}$ | `a\sphat` | $a\sphat$ |
| `\tilde{a}` | $\tilde{a}$ | `\Tilde{a}` | $\Tilde{a}$ | `\widetilde{a}` | $\widetilde{a}$ | `a\sptilde` | $a\sptilde$ |
| `\acute{a}` | $\acute{a}$ | `\Acute{a}` | $\Acute{a}$ | | | | |
| `\bar{a}` | $\bar{a}$ | `\Bar{a}` | $\Bar{a}$ | | | | |
| `\breve{a}` | $\breve{a}$ | `\Breve{a}` | $\Breve{a}$ | | | `a\spbreve` | $a\spbreve$ |
| `\check{a}` | $\check{a}$ | `\Check{a}` | $\Check{a}$ | | | `a\spcheck` | $a\spcheck$ |
| `\dot{a}` | $\dot{a}$ | `\Dot{a}` | $\Dot{a}$ | | | `a\spdot` | $a\spdot$ |
| `\ddot{a}` | $\ddot{a}$ | `\Ddot{a}` | $\Ddot{a}$ | | | `a\spddot` | $a\spddot$ |
| `\dddot{a}` | $\dddot{a}$ | | | | | `a\spdddot` | $a\spdddot$ |
| `\ddddot{a}` | $\ddddot{a}$ | | | | | | |
| `\grave{a}` | $\grave{a}$ | `\Grave{a}` | $\Grave{a}$ | `\imath` | $\imath$ | | |
| `\vec{a}` | $\vec{a}$ | `\Vec{a}` | $\Vec{a}$ | `\jmath` | $\jmath$ | | |

`\dddot`, `\ddddot`, and all of the capitalized commands require the amsmath package. The commands in the fourth column require the amsxtra package. Ⓐ

## A.9.2 Math font commands

| Source | Type | Print |
|---|---|---|
| LaTeX | `\mathbf{A}` | $\mathbf{A}$ |
| | `\mathit{A}` | $\mathit{A}$ |
| | `\mathsf{A}` | $\mathsf{A}$ |
| | `\mathrm{A}` | $\mathrm{A}$ |
| | `\mathtt{A}` | $\mathtt{A}$ |
| | `\mathnormal{A}` | $\mathnormal{A}$ |
| | `\mathcal{A}` | $\mathcal{A}$ |
| amssymb | `\mathbb{A}` | $\mathbb{A}$ |
| | `\mathfrak{A}` | $\mathfrak{A}$ |
| amsmath | `\boldsymbol{\alpha}` | $\boldsymbol{\alpha}$ |

APPENDIX

# Text symbol tables

## B.1 Text accents and fonts

### B.1.1 Text accents

| Type | Print | Type | Print | Type | Print |
|---|---|---|---|---|---|
| \`{o} | ò | \'{o} | ó | \"{o} | ö |
| \H{o} | ő | \^{o} | ô | \~{o} | õ |
| \v{o} | ǒ | \u{o} | ŏ | \={o} | ō |
| \b{o} | o̲ | \.{o} | ȯ | \d{o} | ọ |
| \c{o} | o̧ | \r{o} | o̊ | \t{oo} | o͡o |
| \i | ı | | | \j | ȷ |

### B.1.2 Text font commands

| Command with argument | Command declaration | Switch to |
|---|---|---|
| `\textnormal{...}` | `{\normalfont ...}` | document font family |
| `\emph{...}` | `{\em ...}` | *emphasis* |
| `\textrm{...}` | `{\rmfamily ...}` | roman font family |
| `\textsf{...}` | `{\sffamily ...}` | sans serif font family |
| `\texttt{...}` | `{\ttfamily ...}` | `typewriter style font family` |
| `\textup{...}` | `{\upshape ...}` | upright shape |
| `\textit{...}` | `{\itshape ...}` | *italic shape* |
| `\textsl{...}` | `{\slshape ...}` | *slanted shape* |
| `\textsc{...}` | `{\scshape ...}` | SMALL CAPITALS |
| `\textbf{...}` | `{\bfseries ...}` | **bold** |
| `\textmd{...}` | `{\mdseries ...}` | normal weight and width |

### *B.1.3 Text font size changes (LaTeX and AMS)*

| Command | LaTeX sample text | AMS sample text |
|---|---|---|
| `\Tiny` | [not available] | sample text |
| `\tiny` | sample text | sample text |
| `\SMALL` or `\scriptsize` | sample text | sample text |
| `\Small` or `\footnotesize` | sample text | sample text |
| `\small` | sample text | sample text |
| `\normalsize` | sample text | sample text |
| `\large` | sample text | sample text |
| `\Large` | sample text | sample text |
| `\LARGE` | sample text | sample text |
| `\huge` | sample text | sample text |
| `\Huge` | sample text | sample text |

## B.2 Some European characters

| Type | Print | Type | Print | Type | Print |
|---|---|---|---|---|---|
| \aa | å | \O | Ø | \ss | ß |
| \AA | Å | \oe | œ | \SS | SS |
| \ae | æ | \OE | Œ | ?‘ | ¿ |
| \AE | Æ | \l | ł | !‘ | ¡ |
| \o | ø | \L | Ł | | |

## B.3 Additional text symbols

| Type | Print | Type | Print |
|---|---|---|---|
| \# | # | \$ | $ |
| \% | % | \& | & |
| _ | _ | \textasteriskcentered | ∗ |
| \{ | { | \} | } |
| \dag | † | \textbackslash | \ |
| \ddag | ‡ | \textbar | \| |
| \S | § | \textless | < |
| \P | ¶ | \textgreater | > |
| \copyright | © | \textemdash | — |
| \textregistered | ® | \textendash | – |
| \texttrademark | ™ | \textexclamdown | ¡ |
| \pounds | £ | \textquestiondown | ¿ |
| \textvisiblespace | ␣ | \textquotedblleft | “ |
| \textcircled{a} | ⓐ | \textquotedblright | ” |
| \textsuperscript{a} | $^{a}$ | \textquoteleft | ‘ |
| \textasciicircum | ˆ | \textquoteright | ’ |
| \textasciitilde | ˜ | \textbullet | • |
| | | \textperiodcentered | · |

APPENDIX

C

# TeX, LaTeX, and the Internet

Section C.1 discusses how and where to find the LaTeX distribution, LaTeX packages, and the sample files for this book on the Internet.

There are many different implementations of TeX available: Some commercial versions are covered in Section C.2, and some shareware and freeware versions in Section C.3.

TeX user groups (especially TUG, the TeX Users Group) and the AMS play a significant role in promoting and supporting TeX and LaTeX: see Section C.4.

There is a great deal of useful information on the Internet concerning TeX and LaTeX. Some pointers can be found in Section C.5.

Finally, Section C.6 briefly describes how you can make your articles available to other Internet users.

## C.1 Obtaining files from the Internet

You can download files on the Internet from two types of sites:

- FTP (file transfer protocol) sites
- World Wide Web sites

For both, you use a *client* application on your computer to connect to a *server* on another machine. Today, most *Web browsers,* which are designed to connect to Web sites, also handle FTP transfers.

PCs and UNIX computers include an FTP client as part of the system distribution. On a Macintosh, use Fetch, from Dartmouth College:
`http://www.dartmouth.edu/pages/softdev/fetch.html`
or Anarchie Pro, by Peter Lewis:
`http://www.anarchie-pro.com/anarchie/`

## The Comprehensive TeX Archive Network

The Comprehensive TeX Archive Network (CTAN) is the preeminent collection of TeX-related material on the Internet. There are three main CTAN hosts:

- U.S.:
  - FTP address: `ftp://ctan.tug.org/`
  - Web address: `http://www.ctan.org/`
- U.K.:
  - FTP address: `ftp://ftp.tex.ac.uk/`
  - Web address: `http://www.tex.ac.uk/`
- Germany:
  - FTP address: `ftp://ftp.dante.de/`
  - Web address: `http://www.dante.de/`

It is easier to search for a file or package using the Web sites, but if you know exactly what you want and where it is, downloading from an FTP site can be faster.

There are many *mirrors* (exact duplicates) of CTAN—more than 50 as of this writing. To reduce network load, you should try to use a mirror located near you. Retrieve the document `/tex-archive/CTAN.sites` from any CTAN site to get a list of mirrors.

## The LaTeX distribution

The main LaTeX directory on CTAN is `/tex-archive/macros/latex/`. It has a number of subdirectories, including

- `base`—the current LaTeX distribution
- `required`—packages that all LaTeX installations should have, such as the LaTeX tools, Babel, and graphics
- `contrib`—user-contributed packages
- `unpacked`—the LaTeX distribution in a form that can be downloaded and placed directly in your TeX input directory

### *The AMS packages*

To install the AMS packages, all you have to do is add the contents of the following directories (from a CTAN site) to your TEX input directory:

- `/tex-archive/fonts/amsfonts/latex/`
- `/tex-archive/macros/latex/required/amslatex/`

### *Sample files*

The sample files for this book are available from CTAN sites, in the directory `/tex-archive/info/FirstSteps/`

You may also download the directory `FirstSteps` from `ftp://server.maths.umanitoba.ca/pub/gratzer/`

## C.2 Commercial TEX implementations

Some commercial TEX implementations are ***integrated:*** The application provides an editor, DVI viewer (video driver), and printer driver, as well as the TEX engine. Commercial TEX implementations also provide technical support, whereas most freeware and shareware implementations do not. For a novice, this may be an important consideration.

Two popular integrated implementations are

- PCTEX for a PC: `http://www.pctex.com/`
- TEXTURES for a Macintosh: `http://www.bluesky.com/`

Some users prefer a ***nonintegrated*** setup, which allows them to use the editor of their choice and the best tools for viewing and printing available for their platform. These configurations also allow you to run TEX and the printing application in batch mode, which can be useful for automatically creating documents that change frequently (e.g., price lists, schedules). One such package (for the PC) is Y&Y TEX (`http://www.yandy.com/`).

When producing articles with diagrams that contain formulas, it is important that you be able to copy typeset TEX formulas to Adobe Illustrator (or a similar application), so that the formulas look the same in the diagrams as in the typeset article. All of the above packages have this feature.

The AMS maintains a list of commercial implementations—see `http://www.ams.org/tex/commercial-tex-vendors.html`

## C.3 Freeware and shareware implementations

The most popular TEX implementations are emTEX, MiKTEX, and teTEX, all of which are available from CTAN.

The TeXLive CD is distributed free to all TUG members, and is also available from many other user groups. TeXLive includes TeX implementations for several UNIX variants, AmigaOS, and Windows 95/NT. It also includes a selection of files from the CTAN.

EmTeX is also available on a CD-ROM (4allTeX) from the Dutch user group:

Web site: `http://www.ntg.nl/4allcd/4alltex.html`
E-mail: `ntg@nic.surfnet.nl`

The most popular shareware implementation on the Macintosh is OzTeX: `http://www.kagi.com/akt/oztex.html`

Most Linux distributions include teTeX as an optional installation.

The AMS has a full list of freeware and shareware TeX implementations—see `http://www.ams.org/tex/public-domain-tex.html`

## C.4 TeX user groups and the AMS

There are many user groups around the world that encourage and help people to use and support TeX and LaTeX.

### The TeX Users Group

The TeX Users Group (TUG) does a tremendous job of supporting and promoting TeX, publishing a quarterly journal (*TUGboat*), and organizing an annual international conference. TUG also helps support the LaTeX3 team in maintaining LaTeX and developing LaTeX3.

You should consider joining TUG if you have an interest in TeX or LaTeX. TUG's contact information is:

1466 NW Naito Parkway
Suite 3141
Portland, OR 97209–2820

Telephone: (503) 223-9994
E-mail: `office@tug.org`
Web site: `http://www.tug.org/`

### International TeX user groups

There are also many TeX user groups that are geographic or linguistic in nature. Some of the main groups include:

- Dante (Germany)
- GUTenberg (France)
- NTG (Netherlands)
- UK TUG (United Kingdom)

Links to all of these groups, and many more, can be found on TUG's Web site, at
`http://www.tug.org/lugs.html`

### The American Mathematical Society

The AMS provides excellent technical advice for the AMS packages and AMSFonts. You can reach the AMS technical staff by e-mail at `tech-support@ams.org`, or by telephone at (800) 321-4267, extension 4080, or (401) 455-4080. You will also find a great deal of helpful TeX information on the AMS Web site:
`http://www.ams.org/tex/`

## C.5 Some useful sources of LaTeX information

You may find the Frequently Asked Questions (FAQ) documents maintained on CTAN (in the `/tex-archive/help/` directory) useful. The UK TeX Users Group maintains an FAQ list at
`http://www.tex.ac.uk/cgi-bin/texfaq2html?introduction=yes`
You can also ask most TeX-related questions in the Usenet newsgroup `comp.text.tex`.

Other useful places to start browsing include:

- LaTeX Navigator
  `http://www.loria.fr/services/tex/` (French)
  `http://www.loria.fr/services/tex/english/` (English)
  `http://www.loria.fr/services/tex/german/` (German)
- Sebastian Rahtz's "Interesting TeX-related URLs"
  `http://www.tug.org/interest.html`
- CTAN Catalogue directory
  `/tex-archive/help/Catalogue/ctfull.html`

All of these sites contain many links to other useful sites. The last is a Web index of many of the packages that have been developed for use with LaTeX.

## C.6 Sharing your work via the Internet

Many people choose to make their articles available to others via the Internet, usually through links from their Web pages.

For articles typeset using the Computer Modern fonts and without any graphics, the easiest approach is to put the DVI file on your Web site. If you include graphics in your document, this approach is less useful, because all of the EPS files will need to be downloaded along with the DVI file. There is an additional difficulty: EPS files do not automatically include fonts. You can get around this limitation by converting all text to "outlines," which makes the text into drawings.

By using the PostScript format (see Section 6.4), you can create a single file that contains your article, all of your graphics, and (optionally) the fonts used. Unfortunately, PostScript (PS) files tend to be large, and there are also legal issues that may prohibit you from including commercial fonts in your document.

A better solution is to use Portable Document Format (PDF) files. Adobe's Acrobat Distiller application converts PostScript files to PDF files, which are much smaller (for instance, a PS file of this little book is 9.7 megabytes, but the PDF file is only a little more than a megabyte). When downloaded, this file can be read using the Acrobat Reader application, which is available free of charge for PCs, Macintoshes, and many UNIX systems from Adobe's Web site:
`http://www.adobe.com/prodindex/acrobat/readstep.html#reader`

Software is being developed to obtain PDF files directly from LaTeX.

# *Index*

The Ⓐ symbol indicates an AMS enhancement to LaTeX.

## B

## G



## H



## I



## J

## O



## P

## Q



## R

## T

## U



## V



## W

## Y